Elements

CHLORINE

FLUORINE, BROMINE AND IODINE

Cl F

Br I

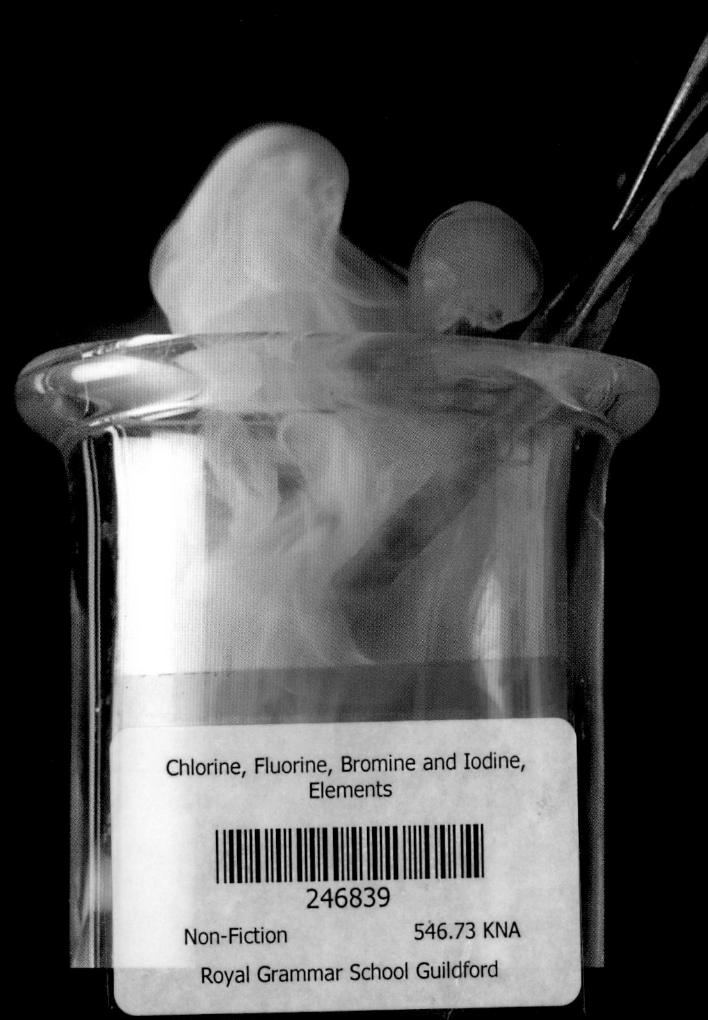

Atlantic Europe Publishing

How to use this book

This book has been carefully developed to help you understand the chemistry of the elements. In it you will find a systematic and comprehensive coverage of the basic qualities of each element. Each two-page entry contains information at various levels of technical content and language, along with definitions of useful technical terms, as shown in the thumbnail diagram to the right. There is a comprehensive glossary of technical terms at the back of the book, along with an extensive index, key facts, an explanation of the Periodic Table, and a description of how to interpret chemical equations.

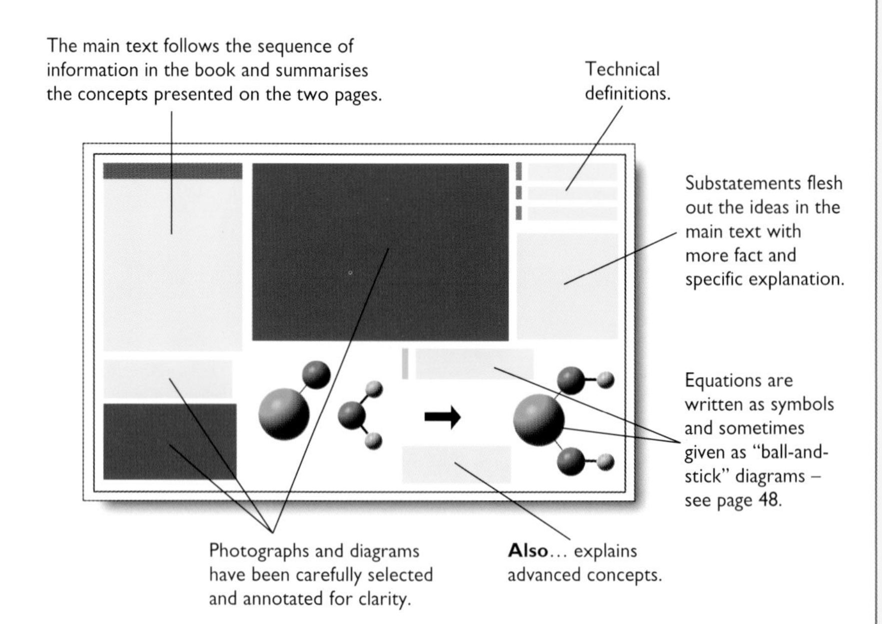

An Atlantic Europe Publishing Book

Author
Brian Knapp, BSc, PhD
Project consultant
Keith B. Walshaw, MA, BSc, DPhil
(Head of Chemistry, Leighton Park School)
Industrial consultant
Jack Brettle, BSc, PhD (Chief Research Scientist, Pilkington plc)
Art Director
Duncan McCrae, BSc
Editor
Elizabeth Walker, BA
Special photography
Ian Gledhill
Illustrations
David Woodroffe
Electronic page make-up
Julie James Graphic Design
Designed and produced by
EARTHSCAPE EDITIONS
Print consultants
Landmark Production Consultants Ltd
Reproduced by
Leo Reprographics
Printed and bound by
Paramount Printing Company Ltd

First published in 1996 by
Atlantic Europe Publishing Company Limited, Greys Court Farm,
Greys Court, Henley-on-Thames, Oxon, RG9 4PG, UK.

Suggested cataloguing location
Knapp, Brian
Chlorine, Fluorine, Bromine and Iodine
ISBN 1 869860 74 8
– *Elements* series
540

Acknowledgements
The publishers would like to thank the following for their kind help and advice: *Jonathan Frankel of* J.M. Frankel and Associates, *Peter Johnson, Irene Knapp, Paul and Molly Stratton* and *ICI (UK).*

Picture credits
All photographs are from the **Earthscape Editions** photolibrary except the following:
(c=centre t=top b=bottom l=left r=right)
Ian Gledhill 43cr; courtesy of **ICI** 18cl, 18tr, 20tr, 39br; **Popperfoto** 26br, 27tl and **SPL** 38tr

Front cover: From left to right the flasks contain: yellow chlorine gas, brown bromine gas and purple iodine vapour.
Title page: Finely divided copper (Dutch metal) spontaneously combusts when put in chlorine gas, producing thick clouds of copper chloride smoke.

Contents

Introduction

An element is a substance that cannot be broken down into a simpler substance by any known means. Each of the 92 naturally occurring elements is therefore one of the fundamental materials from which everything in the Universe is made.

The halogens

The elements that together we call the halogens can all be found in the sea. The ancient Greeks knew where to find them because the word "halogen" comes from the Greek word meaning "salt-producing". Although you cannot see any trace of salt, taste sea water and you will find that it is salty, for dissolved in the water is sodium chloride, a compound of chlorine, one of the halogens, and sodium. Sodium chloride is commonly known as table salt.

Sodium chloride is not the only salt in sea water. You will also find salts containing fluorine (as in the fluoride put into your toothpaste), iodine (as in the substance used to iodise table salt) and bromine (as in the bromide used on photographic film).

The sea contains natural collectors of some of these valuable elements. In particular, iodine is gathered by seaweed and stored in its tissues.

Each of the halogen elements is so reactive you will never find a halogen alone. Halogens bind so well to other atoms that great energy has to be applied if they are to be separated from their compounds. This may be just as well, because all of the halogen elements are very dangerous. Chlorine, for example, has been used as a chemical weapon in wartime, and fluorine gas is so hazardous that laboratory demonstrations using it were not even attempted for this book!

As compounds, though, the halogens are harmless, even beneficial. Think of the halogens as health-givers. Iodine salts are essential to prevent thyroid gland problems, fluoride strengthens your teeth, and chlorine salts can be used to preserve food and disinfect water supplies. Halogens also make up many of the commonly used anaesthetics used in surgery. Chloroform (a compound containing chlorine) was one of the early anaesthetics; halothane (a compound containing fluorine, chlorine and bromine) is now used in its place.

◀ The use of chloroform provided a major step forward in anaesthetics and allowed surgery to be a less traumatic experience than had previously been the case. Procaine hydrochloride (novocaine) is one of the most widely used modern local anaesthetics, used extensively by dental surgeons. Like other local anaesthetics, procaine blocks nerve impulses that carry pain signals.

The halogens

The halogens – fluorine, chlorine, bromine and iodine – occur in the same group of the periodic table and therefore have many characteristics in common. Some of these are shown on these two pages.

▼ (left) Chlorine gas; (middle) bromine vapour over bromine liquid, (right) iodine vapour over solid crystals of iodine (iodine does not occur as a liquid). Note: Fluorine is *so* dangerous, it is *never* used in school and college demonstrations and so is not shown here.

Also... Why halogens are so reactive

Each of the elements in the halogen family contains seven electrons in its outermost electron shell. To be stable it needs to have eight. As a result the atoms react with other elements to gain an electron and fill the shell. The smallest atoms try to fill their electron shell most vigorously, and as fluorine is the smallest of the family, it is the most reactive. Iodine has the largest atoms and is the least reactive of the family (see pages 44 and 45).

Light causes halogens to decompose and change colour

These two test tubes below right show white silver chloride shortly after it has been prepared as a precipitate, and the same tube after it has been exposed to light for a few minutes and has darkened. (These changes are the basis of how a photographic film works. Silver bromide is used as it is the most sensitive to light.)

decompose: to break down a substance (for example by heat or with the aid of a catalyst) into simpler components. In such a chemical reaction only one substance is involved.

organic substance: a substance that contains carbon.

solvent: the main substance in a solution (e.g. water in salt water).

EQUATION: Silver halides and sunlight

Silver chloride ➪ silver + chlorine

$2AgCl(s) \Rightarrow 2Ag(s) + Cl_2(g)$

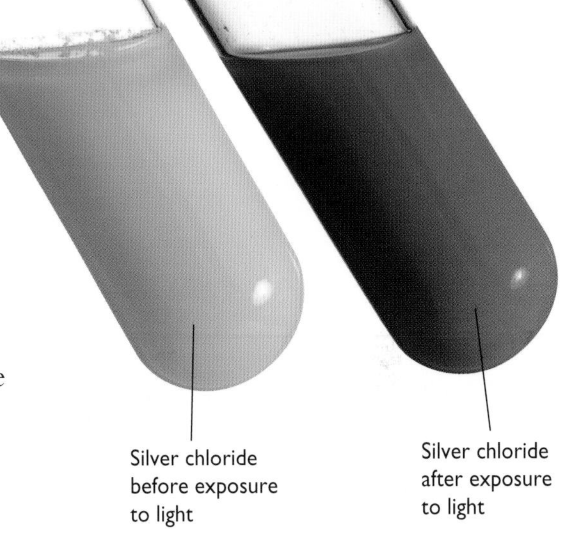

Halogens dissolve better in organic solvents than in water

In the photographs below, iodine, bromine and chlorine have been introduced to a tube containing two liquids in which other substances commonly dissolve (solvents). The bottom one is water and above it is an organic liquid (methyl benzene). Each of the halogens is more soluble in the organic solvent, and it has turned the liquid dark brown. They are barely soluble in water, and so the water is only slightly coloured.

▶ Halogens have been dissolved in methyl benzene.

Halide minerals

Most halides – compounds containing chlorine, fluorine, iodine or bromine – readily dissolve in water. They are only precipitated when water is evaporated, such as happens when desert lakes dry up or in coastal lagoons. The precipitates form white crystals, most commonly seen as the white "ghost lakes" or playas in deserts.

White deposits also form on the surface of some soils, especially those that have been irrigated with water containing high concentrations of dissolved halides. Soils that exhibit white surface encrustations are contaminated with halides to a level where they are toxic to most plants. Such soils are examples of desertification.

▲ Fluorine is found as fluorite (calcium fluoride), a blue, yellow or green crystalline mineral. Often known as Blue John, this mineral makes beautiful cubic (box-shaped) crystals.

▲ Each time a wave breaks, tiny droplets of water spray are thrown into the air. Many of these are carried aloft by the wind, where the sodium chloride they contain attracts even more water. These tiny particles are called condensation nuclei by meteorologists because they cause water to condense on them, gradually growing bigger and bigger until they reach a size at which they can fall out of the clouds as rain. This is why many freshwater raindrops contain tiny amounts of salt.

► The glassy character of roc salt (halite) can clearly be seen in this specimen.

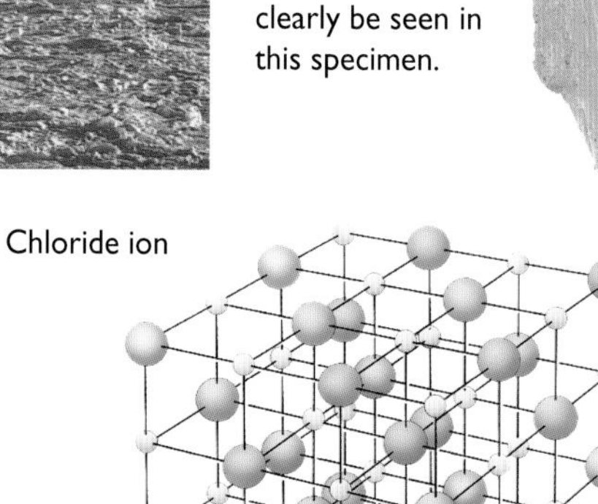

◄ The structure of halite (sodium chloride) is made by sodium and chloride ions occupying alternate corners of a cubic lattice.

Fluorite

Fluorite is calcium fluoride, a mineral that exists as beautiful cubic (box-shaped) crystals in greens, blues and yellows. It is a relatively soft mineral that can be scratched with a knife. It is commonly found in thick salt beds and also in veins and near hot springs. One of the most interesting properties of fluorite is that it glows – fluoresces – in ultraviolet light.

▶ This seaweed is a natural scavenger of iodine from sea water. Coastal farming communities often use it as a natural fertiliser.

Halite

Common salt, or sodium chloride, forms thick beds. Sometimes it occurs as crystals. The crystals of the mineral, called halite or rock salt, are cubic (look closely at table salt to check this), glassy or white, but may also look pink because of iron staining. Halite is very soft and can be scratched by a copper coin.

Salt makes up four fifths of all the solids dissolved in the sea. Blood is also a salt (saline) solution in which red and white blood cells are suspended. This is why, when people lose blood, doctors often administer a saline solution until a matching blood supply can be found.

The body uses the sodium ions for controlling osmosis and the chloride ions to control the nerves.

chloride: a substance containing chlorine.

desertification: a process whereby a soil is allowed to become degraded to a state in which crops can no longer grow, i.e. desert-like. Chemical desertification is usually the result of contamination with halides because of poor irrigation practices.

fluorescent: a substance that gives out visible light when struck by invisible waves such as ultraviolet rays.

halide: a salt of one of the halogens (fluorine, chlorine, bromine and iodine).

halite: the mineral made of sodium chloride.

osmosis: a process where molecules of a liquid solvent move through a membrane (filter) from a region of low concentration to a region of high concentration of solute.

playa: a dried-up lake bed that is covered with salt deposits. From the Spanish word for beach.

saline: a solution in which most of the dissolved matter is sodium chloride (common salt).

Iodine and bromine

Iodine and bromine do not make rock-forming minerals. Instead, they are usually all found in sea water. Some kinds of seaweed and molluscs are naturally rich in iodine and can be used as sources of the element.

▲ Playa lakes are salt beds that form when temporary desert lakes dry up. The majority of these deposits are formed of sodium chloride, common salt that has been dissolved from rocks by rainfall and carried to lakes in the floods that follow torrential desert downpours. Playa lakes demonstrate the amount of salt that is carried in even "fresh" water.

Chlorine

Chlorine is a greenish-yellow gas (the name chlorine in Greek means greenish-yellow), which is more dense than air and which will dissolve to some extent in water. It forms an amber-coloured liquid at its boiling point.

Nearly one-fiftieth of sea water is made of chlorine, where it occurs combined with sodium as sodium chloride.

People become sensitive to the presence of chlorine in concentrations as little as three to five parts per million because it irritates the membranes of the nose. At concentrations of 30 parts per million its effects on the eyes and nose become severe. At only slightly higher concentrations, it causes difficulty in breathing.

Like other halogens, chlorine is a very reactive substance and it is never found as a free element in nature. It is particularly reactive in the presence of heat, and although it does not dissolve well in water, it does react in the presence of moisture. It is particularly reactive with organic substances, sometimes causing explosions. For these reasons chlorine needs to be kept in dry, cool conditions.

A flame will continue to burn in a vessel containing chlorine gas. Many products containing chlorine (such as PVC, polyvinyl chloride) will decompose when heated. If used extensively for home furnishings, for example, these materials can be a major hazard during a fire.

Chlorine is one of the ten most important industrial chemicals. The success of an industrial country is sometimes measured in the amount of chlorine its chemical factories use!

Laboratory preparation of chlorine

A gas jar of chlorine gas is prepared by reacting concentrated hydrochloric acid on potassium permanganate crystals. Chlorine is heavier than air, so it can be allowed to collect at the bottom of a gas jar, forcing the air out of the top.

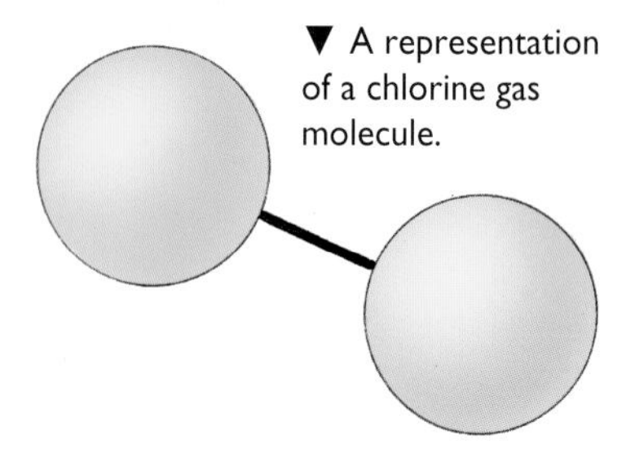

▼ A representation of a chlorine gas molecule.

Dangers of chlorine

Transporting chlorine can be a health hazard. If the transporting vessel is involved in an accident and the chlorine escapes to the atmosphere, many people might have their lungs damaged or even be killed. For this reason chlorine is not usually transported; instead other chemicals are brought to where the chlorine is made. This is the case, for example, with making the common plastic PVC. A material called ethene is brought to the chlorine plant to be reacted with the chlorine. The product is shipped off to be further manufactured elsewhere.

Also...

When large amounts of waste hydrochloric acid are available – for example, when PVC is being made – the chlorine can often be recovered and recycled by oxidising the acid.

chlorination: adding chlorine to a substance.

oxidation: a reaction in which the oxidising agent removes electrons. (Note that oxidising agents do not have to contain oxygen.)

EQUATION: Preparation of chlorine

Concentrated hydrochloric acid + potassium permanganate ⇨ chlorine + water + manganese chloride + potassium chloride

$16HCl(aq) + 2KMnO_4(s) \Rightarrow 5Cl_2(g) + 8H_2O(l) + 2MnCl_2(aq) + 2KCl(aq)$

Chlorine gas is collected in a gas jar. It is heavier than air.

Chlorine and combustion

Chlorine is a very reactive element because it is an oxidising agent. In other words, materials will burn (combust) easily in chlorine. This can be an unusual sight because we are so used to materials burning in air. Furthermore, some materials actually catch fire when they are placed in chlorine and the reaction generates a large amount of heat (it is an exothermic reaction).

EQUATION: Chlorine reacts with finely divided copper

Chlorine + copper ⇨ copper chloride

$Cl_2(g) + Cu(s) \Rightarrow CuCl_2(s)$

Chlorine

Copper

❶▼ Finely divided copper (Dutch metal) spontaneously catches fire when put in chlorine, showing that chlorine is an oxidising agent.

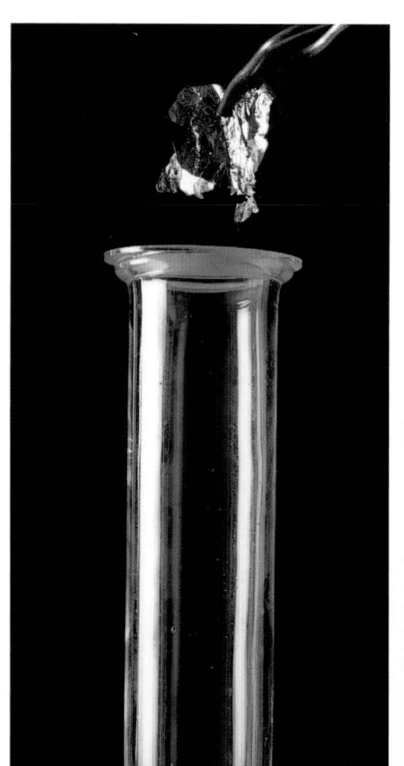

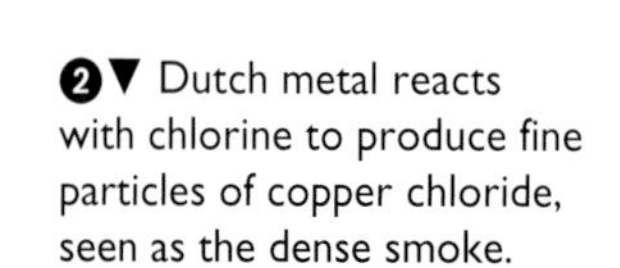

❷▼ Dutch metal reacts with chlorine to produce fine particles of copper chloride, seen as the dense smoke.

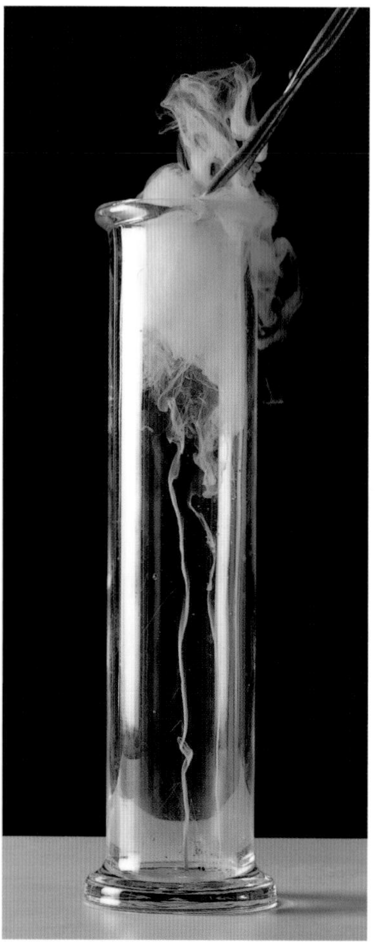

❸▼ The smoke and gas fill the gas jar.

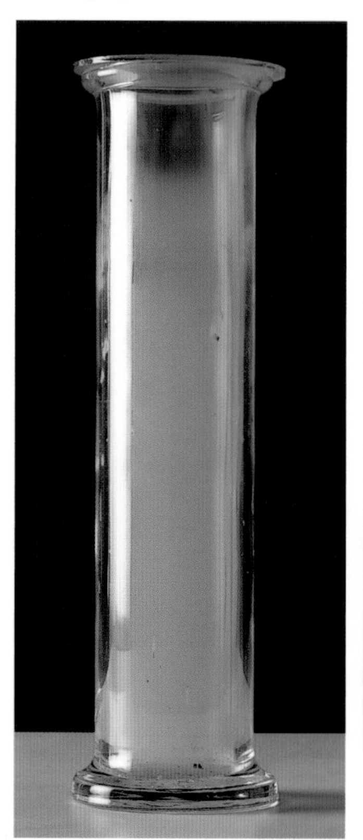

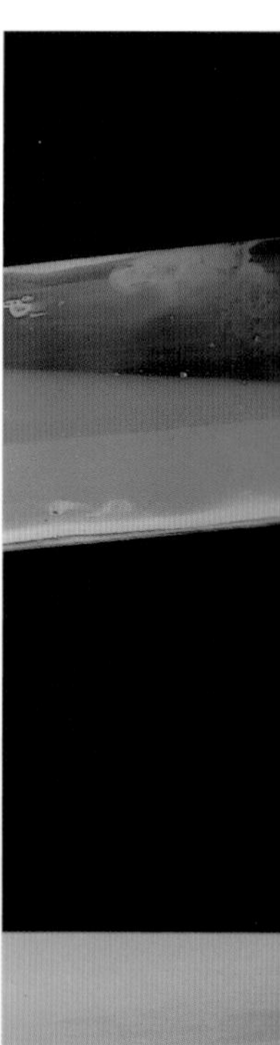

combustion: the special case of oxidisation of a substance where a considerable amount of heat and usually light are given out. Combustion is often referred to as "burning".

exothermic reaction: a reaction that gives heat to the surroundings. Many oxidation reactions, for example, give out heat.

oxidation: a reaction in which the oxidising agent removes electrons. (Note that oxidising agents do not have to contain oxygen.)

smoke: a mixture of both small solids and gases.

◀▶ In the picture on the left a wax taper is being held in the air above a gas jar of chlorine. The taper burns in the oxygen in the air with a yellow flame, releasing carbon (soot) particles and forming carbon dioxide gas.

However, when the wax taper is introduced to a gas jar of chlorine as shown in the picture on the right, the taper burns with a red flame. It also liberates carbon particles (soot) and produces a steamy gas – highly poisonous hydrogen chloride – that attracts water vapour in the air to form tiny droplets of hydrochloric acid. This is not a nice gas to be near!

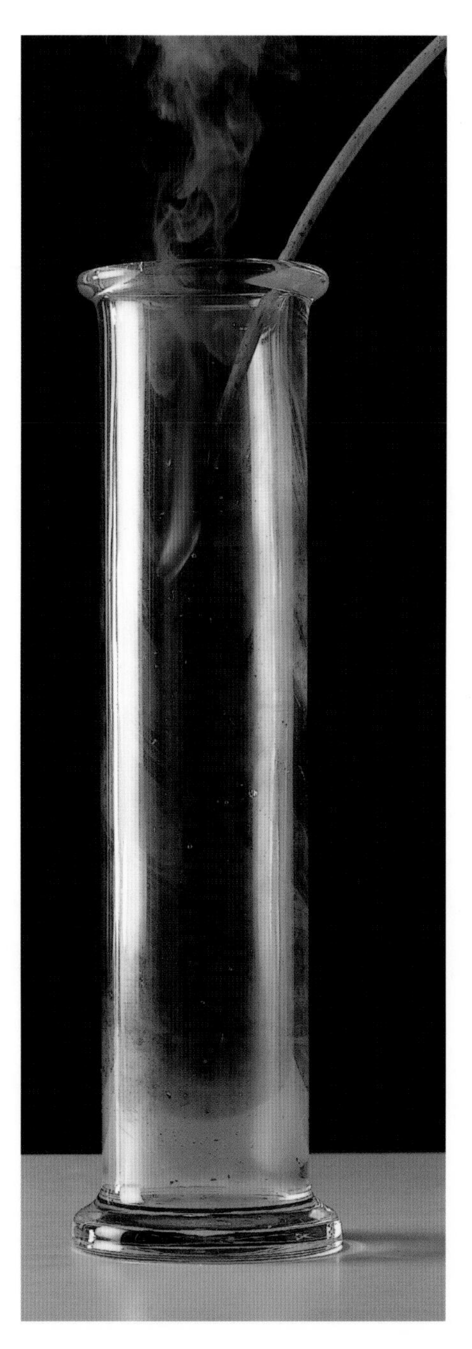

❹◀ Chlorine, containing copper chloride smoke, is denser than air and can therefore be poured from a gas jar.

Chlorine as a bleaching agent

A bleach is a substance that removes stains from materials. It also has a disinfectant action.

There are two kinds of bleach: those that oxidise and those that reduce substances. Chlorine-based bleaches are oxidising agents.

Domestic bleach was traditionally made by passing chlorine gas over dry calcium oxide, producing calcium chlorate. When put in water this substance formed domestic bleach.

Powdered calcium chlorate, which can be added to water, and liquid solutions of sodium chlorate are both sold as domestic bleach (see page 24). In either case the compounds produced are quick and cheap to make, which accounts for the low price of most domestic bleaches.

Bleaches containing chlorine are powerful and are suited only to some fabrics, such as cotton, linen, and synthetic fibres. They cannot be used with wool, silk and some other materials, so in general chlorine-based bleaches are not used in detergent powders.

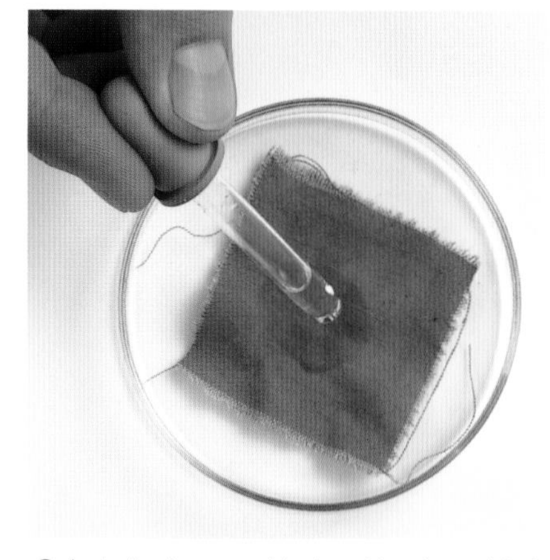

❶▲ A dyed material before bleach is added.

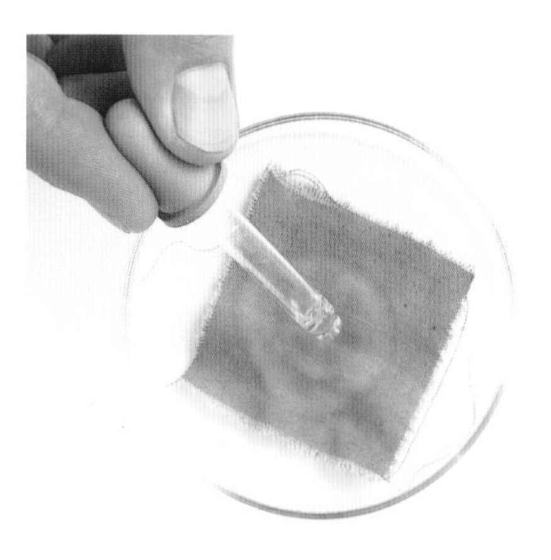

❷▲ The bleach reacts instantaneously with the dye.

EQUATION: The reaction of chlorine and sodium hydroxide to make liquid bleach

Chlorine + sodium hydroxide ⇨ sodium chlorate + water + sodium chloride

$Cl_2(g) + 2NaOH(aq) \Rightarrow NaOCl(aq) + H_2O(l) + NaCl(aq)$

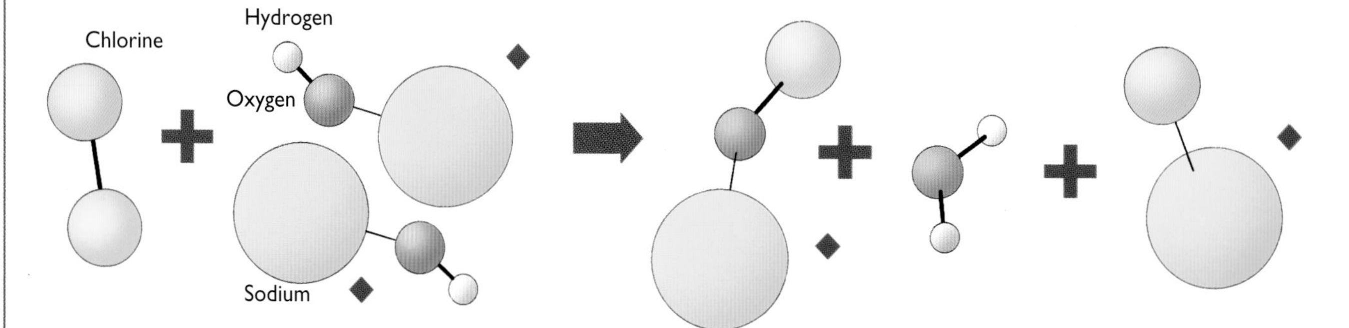

Demonstrating the bleaching action of chlorine

The bleaching action of chlorine gas can be seen by bubbling chlorine through a solution containing the vegetable dye litmus, an indicator. The sequence of pictures shows the progressively lighter colour of the solution.

bleach: a substance that removes stains from materials either by oxidising or reducing the staining compound.

indicator: a substance or mixture of substances that change colour with acidity or alkalinity.

oxidising agent: a substance that removes electrons from another substance (and therefore is itself reduced).

reducing agent: a substance that gives electrons to another substance.

❶▲ Chlorine gas is produced by reacting hydrochloric acid and potassium manganate.

❷ ▲ Chlorine is bubbled through the gas jar containing litmus dye.

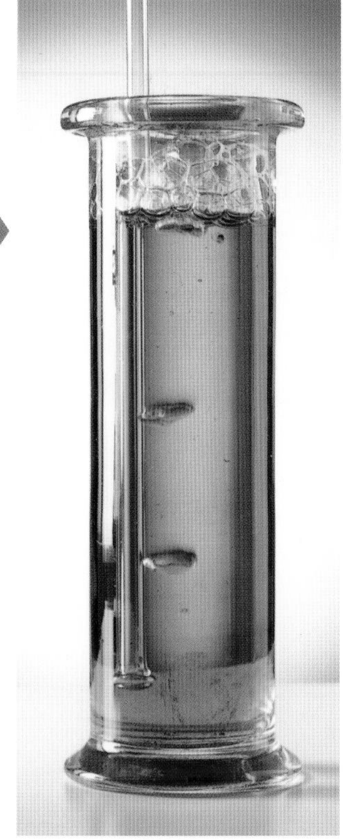

❸ ▲ The solution gradually clears as the dye is bleached.

How bleach works

A coloured natural material gets its colour from the way that certain combinations of carbon atoms are linked together by double bonds. Synthetic dyes are mainly also based on carbon compounds.

The colouring chemicals are able to absorb some forms of light rays while reflecting others. When a bleach reacts with the natural material, the bleach breaks up the bonds, forming new substances that are not able to absorb light and so are colourless.

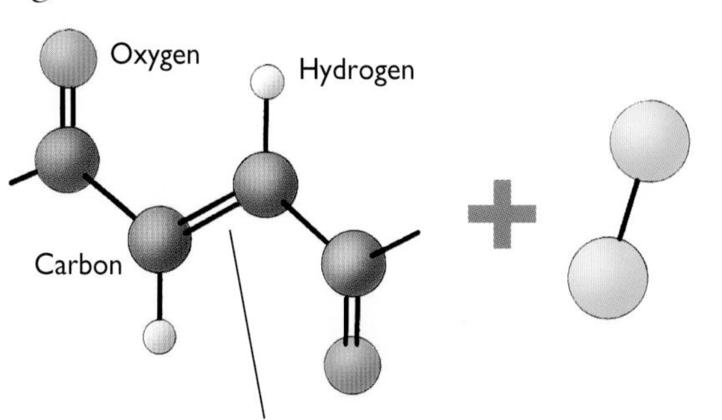

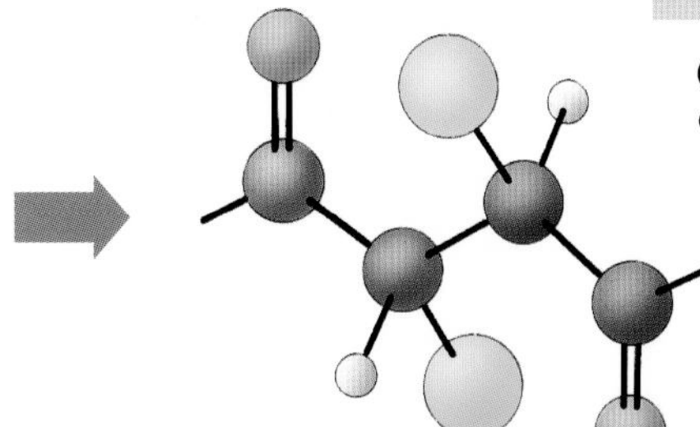

Colour in this solution is due to the two bonds here. The oxidising effect of chlorine breaks these bonds and bleaches the solution.

Chlorine as a disinfectant

Disinfecting is the process of killing bacteria and microorganisms. Chlorine, often in the form of calcium chlorate powder, is an effective way of disinfecting water supplies, swimming pools and sewage systems.

The use of chlorine dates back to the last century, when most large cities had developed sewage systems that simply fed raw sewage into rivers. The same rivers were also used for drinking water. Even in places where water was obtained from wells, it was not safe to drink because of sewage seeping from nearby cess pits into the underground water supplies.

An infamous cholera epidemic occurred in London in 1848 that revolutionised the way people thought about disinfecting water supplies. Cholera caused over 25,000 deaths in the following decade. Nothing could be done about it because of the lack of sewage treatment. The nearby River Thames received most of the raw sewage and produced a foul smell. In 1855, as a result of the research done into the epidemic and the smell (which proved to have the same cause), London became the first city in the world to have a sewage treatment system based on the use of chlorine.

The disinfectant properties of chlorine were also needed in hospitals. When surgeons began to disinfect their hands before each operation, the number of fatalities decreased rapidly. The introduction of such disinfectants is closely connected with the efforts of the famous nurse Florence Nightingale.

Water supplies

Clean water supplies are essential for healthy living. The water entering a treatment plant contains chemical pollutants, small particles of solids, and biological material such as bacteria and algae.

The first stages of treatment involve filtering out the solid particles. The remaining chemicals are neutralised and precipitated. The water is then treated with chlorine, which is used as a disinfectant. Chlorine also oxidises any remaining organic materials.

▼ Swimming pools that get intensive use, such as at tropical resorts, often smell strongly of the chlorine that has been added as disinfectant. The chlorine can irritate sensitive tissues such as those in the nose and eyes.

◀ Part of a water treatment plant.

disinfectant: a chemical that kills bacteria and other microorganisms.

neutralisation: the reaction of acids and bases to produce a salt and water. The reaction causes hydrogen from the acid and hydroxide from the base to be changed to water. For example, hydrochloric acid reacts with sodium hydroxide to form common salt and water. The term is more generally used for any reaction where the pH changes towards 7.0, which is the pH of a neutral solution.

precipitate: tiny solid particles formed as a result of a chemical reaction between two liquids or gases.

EQUATION: The reaction of chlorine and water

Water + chlorine ➪ hypochlorous acid + hydrochloric acid

$H_2O(l)$ + $Cl_2(g)$ ➪ $HOCl(aq)$ + $HCl(aq)$

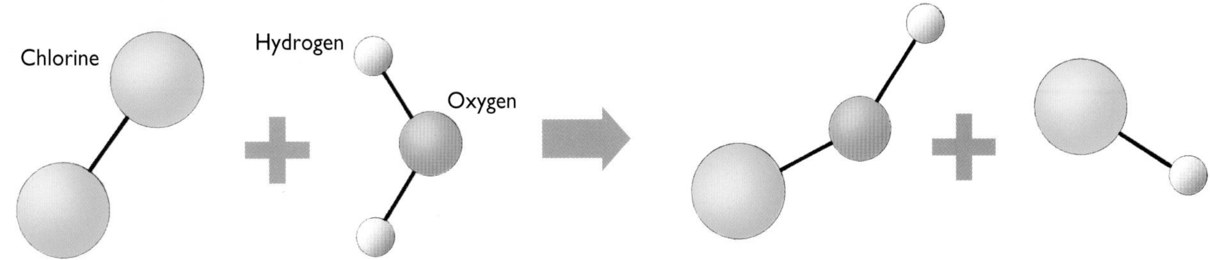

The chlorination of swimming pools

Disinfectants are needed in swimming pool water to prevent the growth of bacteria and algae that might eventually become a health risk.

Chlorine gas was the disinfectant traditionally used in public baths. It was also sold as a solid compound for domestic swimming pools. The chlorine reacts with the water to form hypochlorous acid, which in turn kills bacteria and algae.

Adding more and more chlorine causes the acidity of the water to rise, and this can have harmful side effects, such as making the eyes smart. So to prevent this, a neutralising chemical, sodium carbonate, is added.

The chlorine smell in a swimming pool is not the chlorine itself, but the smell of ammonia-like gases released as the chlorine gets to work on organic matter.

Disinfectants are now available without the side-effects caused by chlorine.

Making chlorine: the mercury cathode cell

Chlorine is one of the main chemicals used in modern plastics manufacture. There are two main ways of producing it.

About three-quarters of the chlorine needed for industry is produced by the mercury cathode cell process described below. Another process, the diaphragm cell process, is described on pages 20 and 21.

▲ Chlorine is stored in bottled form.

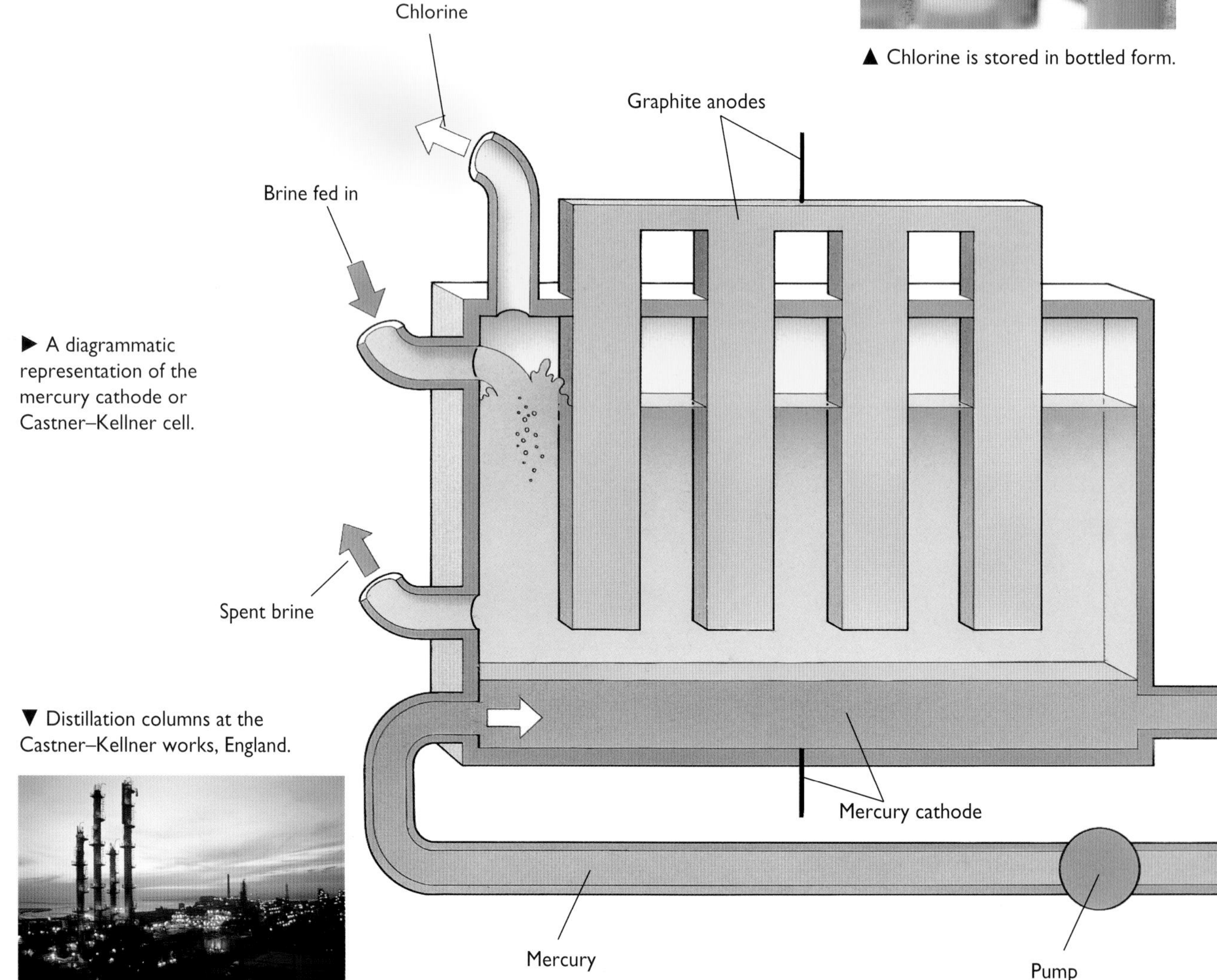

► A diagrammatic representation of the mercury cathode or Castner–Kellner cell.

▼ Distillation columns at the Castner–Kellner works, England.

Operating the mercury cathode cell

The mercury cathode, also known as the Castner–Kellner cell after its inventors, is an electrolytic process using an electric current to dissociate sodium chloride (brine). It consists of a cell with a graphite anode and a bed of mercury as the cathode. Sodium chloride solution is used as the electrolyte.

A huge current (about 300,000 amps – about 6000 times as much as a household might use with all of its appliances switched on!) is passed through the cell. Chlorine is released at the anode, where it is collected.

Mercury is used as the cathode because it readily forms an amalgam with other metals. Sodium is a metal ion and makes an amalgam with the mercury. When the mercury cannot absorb any more sodium, the amalgam is carried away and the sodium extracted and made into sodium hydroxide by reaction with water. The refined mercury is then reused and the sodium hydroxide is sold to help pay for the process.

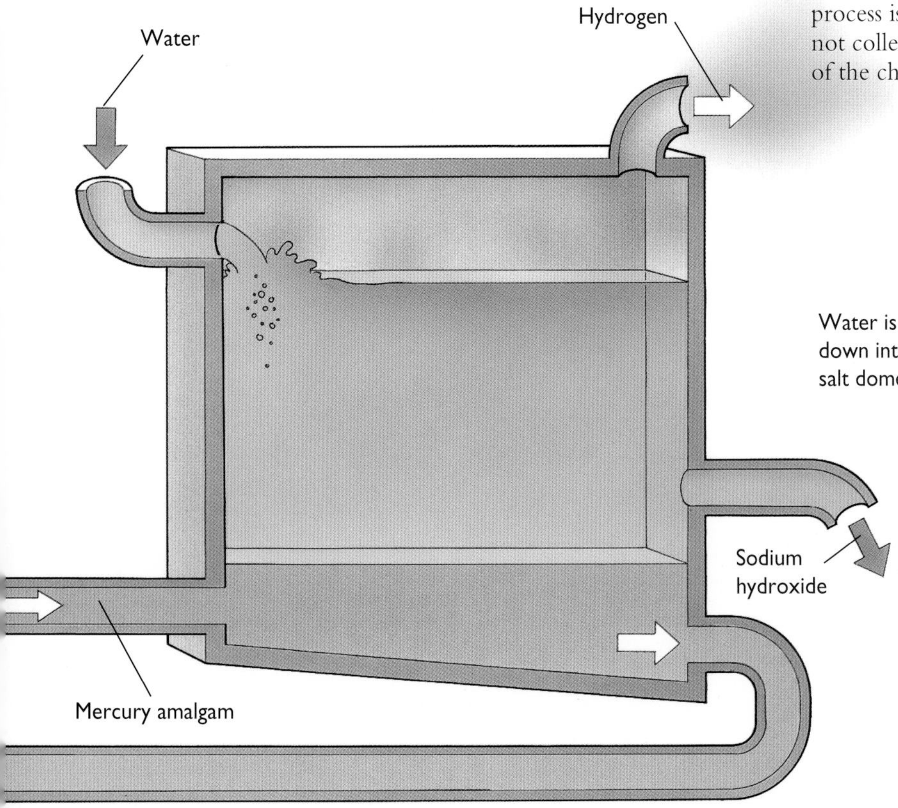

Water
Hydrogen
Sodium hydroxide
Mercury amalgam

amalgam: a liquid alloy of mercury with another metal.

cell: a vessel containing two electrodes and an electrolyte that can act as an electrical conductor.

electrolysis: an electrical–chemical process that uses an electric current to cause the break up of a compound and the movement of metal ions in a solution. The process happens in many natural situations (as for example in rusting) and is also commonly used in industry for purifying (refining) metals or for plating metal objects with a fine, even metal coating.

electrolyte: a solution that conducts electricity.

ion: an atom, or group of atoms, that has gained or lost one or more electrons and so developed an electrical charge. Ions behave differently from electrically neutral atoms and molecules.

Also...

Chlorine can also be produced as a byproduct of making sodium using the Downs process. Salt is mixed with calcium chloride and heated until it is molten. Chlorine gas is given off at the anode of the cell and can be collected for use. The Downs process is designed for sodium refining and does not collect sufficient chlorine for all the demands of the chlorine industry.

▼ The raw material for the production of chlorine is sodium chloride, rock salt, which often occurs in rocks in the form of a salt dome. It can be mined or extracted from the ground by dissolving it in hot water and then pumping it to the surface.

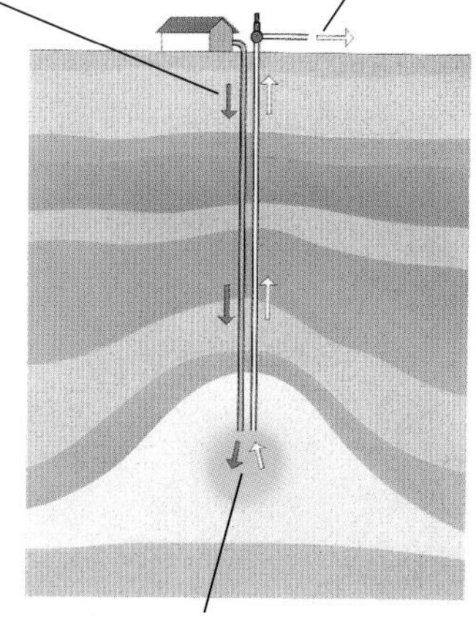

Water is pumped down into the salt dome.

Brine is pumped up to the surface.

Water dissolves rock salt to make a concentrated salt solution called brine. This can then be drawn back up to the surface.

Making chlorine: the diaphragm cell

The diaphragm cell is a piece of equipment that allows brine solution and sodium hydroxide to be used to manufacture a number of chemicals, including chlorine. It is an alternative to the mercury cathode cell process shown on the previous page.

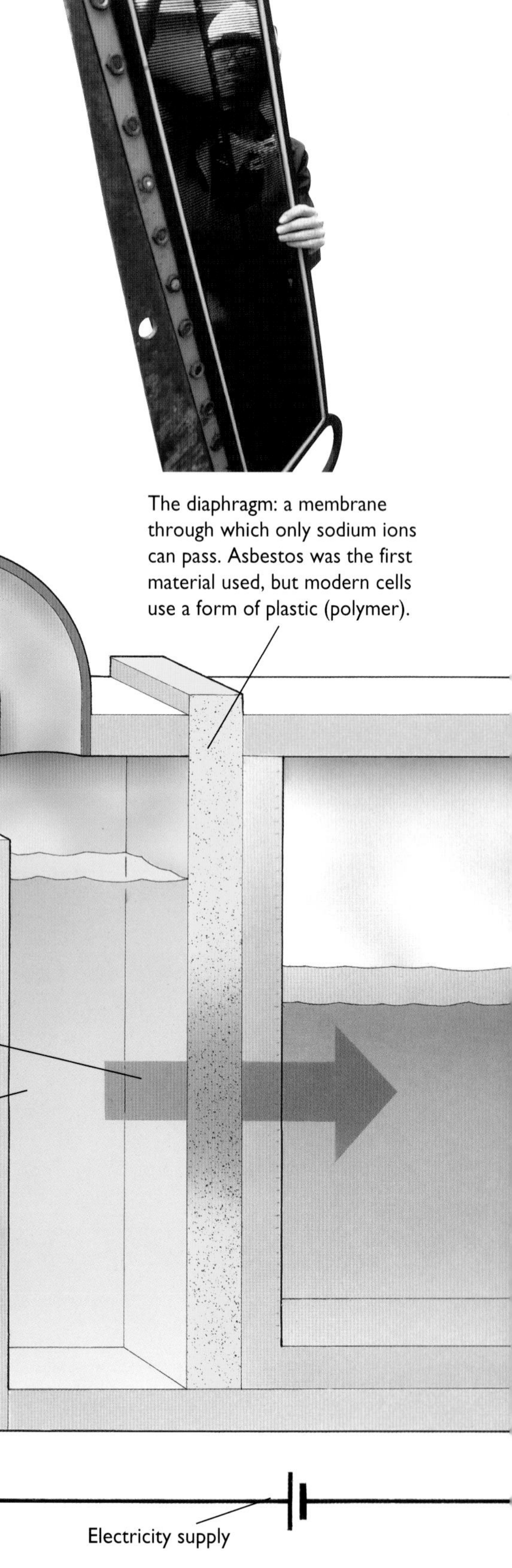

◀ A diaphragm.

▼ A diagrammatic representation of the electrolysis process for manufacturing sodium hydroxide.

alkali: a base in solution.

diaphragm: a semipermeable membrane – a kind of ultra-fine mesh filter – that will allow only small ions to pass through. It is used in the electrolysis of brine.

electrolysis: an electrical–chemical process that uses an electric current to cause the break up of a compound and the movement of metal ions in a solution.

EQUATION: Electrolysis of a salt solution

Sodium chloride + water ⇨ sodium hydroxide + chlorine + hydrogen

$$2NaCl(aq) + 2H_2O(l) \xrightarrow{\text{electrical energy}} 2NaOH(aq) + Cl_2(g) + H_2(g)$$

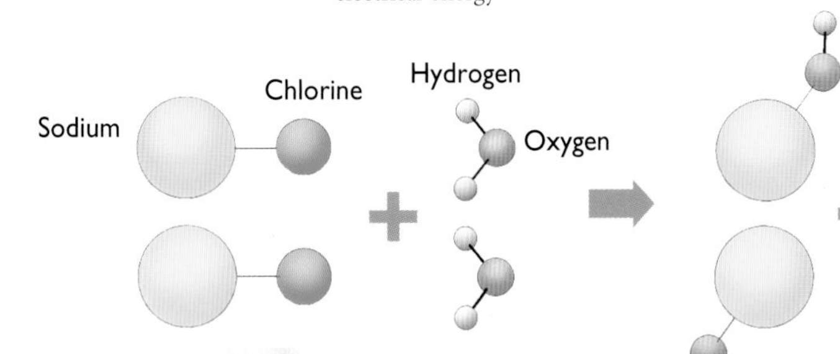

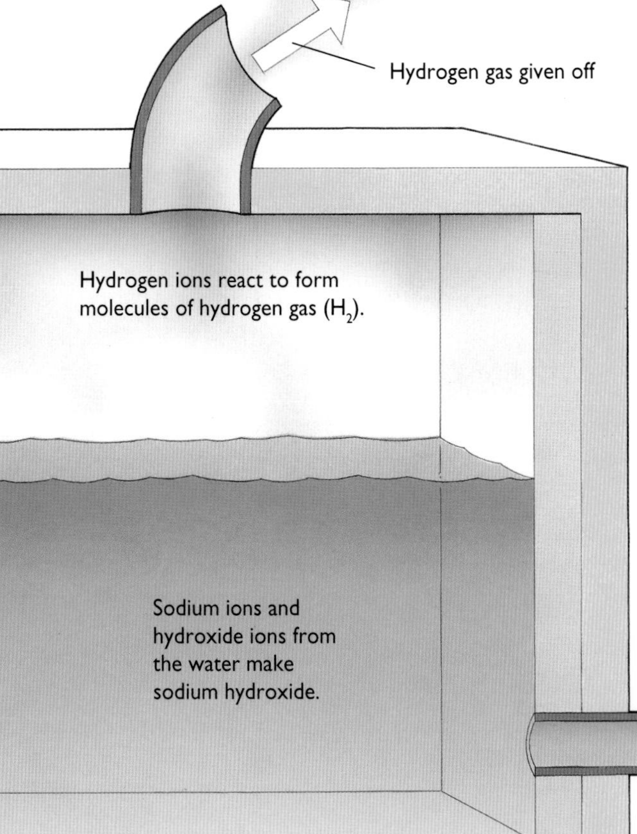

The diaphragm cell

The diaphragm cell is constructed in such a way as to make use of an electrolytic process to collect two gases simultaneously. As with the mercury cathode cell, an electric current is passed through a brine solution, but instead of amalgamating sodium ions with mercury to leave chlorine, the combination of brine and sodium hydroxide creates two useful gases, chlorine and hydrogen.

The cell consists of two parts, each filled with brine, but separated by a porous diaphragm (a kind of filter) made of asbestos. The diaphragm ensures that the hydrogen and chlorine gases can be collected separately.

Electrolysis releases hydrogen gas from the water and chlorine gas from the salt. Chlorine is given off at the positive electrode (the anode) and hydrogen is given off at the negative electrode (the cathode). The process dilutes the brine and strengthens the sodium hydroxide solution. Sodium hydroxide is a strong alkali and a valuable byproduct.

Halogens in plastics

One of the main uses of the halogens (and of chlorine in particular) is in the production of plastics.

One of the simplest procedures is to react products of an oil refinery (such as the gas ethene) with chlorine or fluorine.

Polyvinyl chloride (known as vinyl, polychloroethene or PVC) is made with ethene and chlorine. The reaction that takes place exchanges one chlorine atom for one hydrogen atom of the ethene.

Polyvinyl fluoride (polyfluoroethene or PVF) is made in a similar way using fluorine gas. As with vinyl, a fluorine atom replaces a hydrogen atom. This plastic is heat-resistant.

Polyvinyl chloride

Polyvinyl chloride is a common thermoplastic, meaning a plastic that can be softened and reworked time after time.

Vinyl chloride is made by reacting ethene or acetylene (gases made from petroleum refining) with hydrochloric acid.

In this reaction, one hydrogen atom in ethene is replaced with a chlorine atom. This change produces a gas whose molecules can be linked together (polymerised) to make polyvinyl chloride.

▼ Vinyl chloride is often used as house cladding and window frames.

Using polyvinyl chloride

Polyvinyl chloride is a hard plastic that is rarely used on its own. The material we call vinyl or PVC is actually a mixture of polyvinyl chloride and other compounds that make it soft and flexible.

Although it does not burn, when heated polyvinyl chloride decomposes and releases hydrogen chloride gas, together with carbon monoxide. Both gases are toxic, so this limits the uses for polyvinyl chloride in materials such as furniture fabrics.

The main uses of polyvinyl chloride are in plastic window frames, house cladding, hose pipes, watering cans and waterproof "plastic" sheeting.

The structure of polyvinyl chloride

Polyvinyl chloride is a hydrocarbon chain, incorporating chlorine. The same structure but without the chlorine produces polythene, used to make plastic bags. By substituting chlorine the properties are changed so that the material is more durable and rigid.

hydrocarbon: a compound in which only hydrogen and carbon atoms are present. Most fuels are hydrocarbons, as is the simple plastic polyethene (known as polythene).

plastic (material): a carbon-based material consisting of long chains (polymers) of simple molecules. The word plastic is commonly restricted to synthetic polymers.

polymer: a compound that is made of long chains by combining molecules (called monomers) as repeating units. ("Poly" means many, "mer" means part).

EQUATION: Reaction to produce vinyl chloride (chloroethene)

Ethene gas + chlorine gas ⇨ chloroethene gas + hydrogen chloride

$CH_2{=}CH_2(g) + Cl_2(g) \Rightarrow CH_2{=}CHCl(g) + HCl(g)$

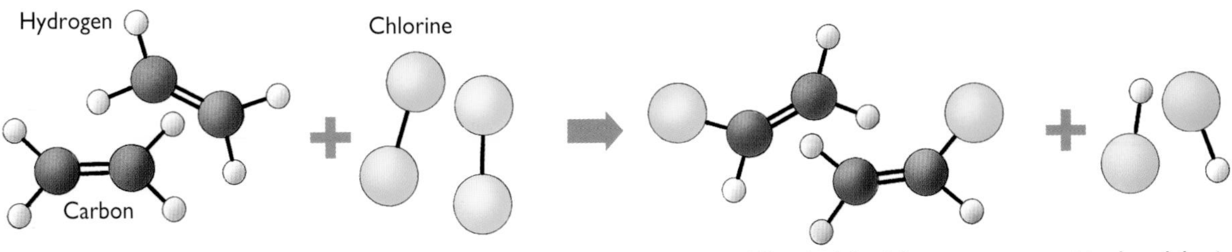

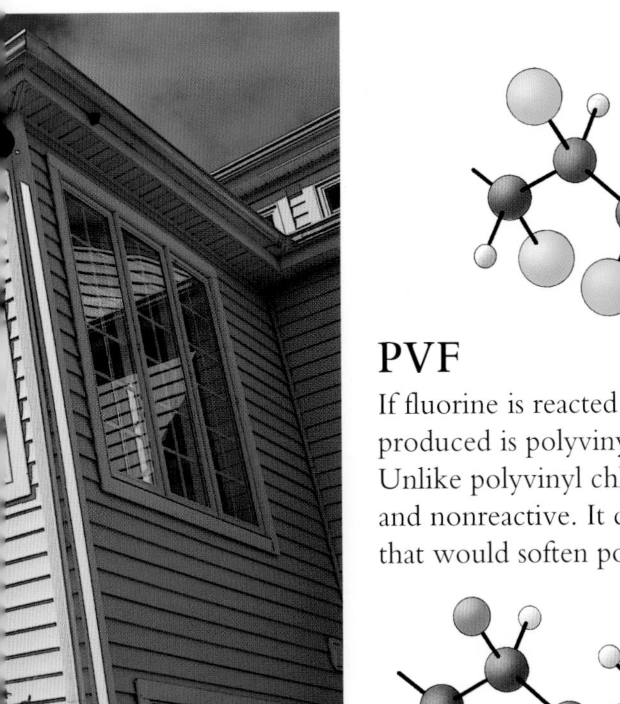

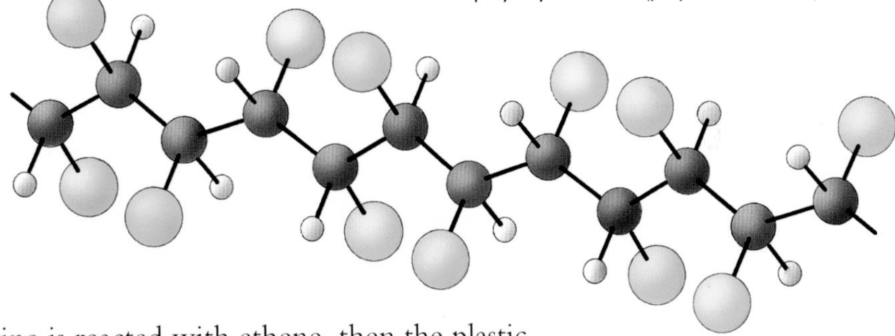

▼ Vinyl chloride polymerises to produce polyvinyl chloride (polychloroethene).

PVF

If fluorine is reacted with ethene, then the plastic produced is polyvinyl fluoride (fluoroethene). Unlike polyvinyl chloride, this plastic is tough and nonreactive. It can be used to hold solvents that would soften polyvinyl chloride.

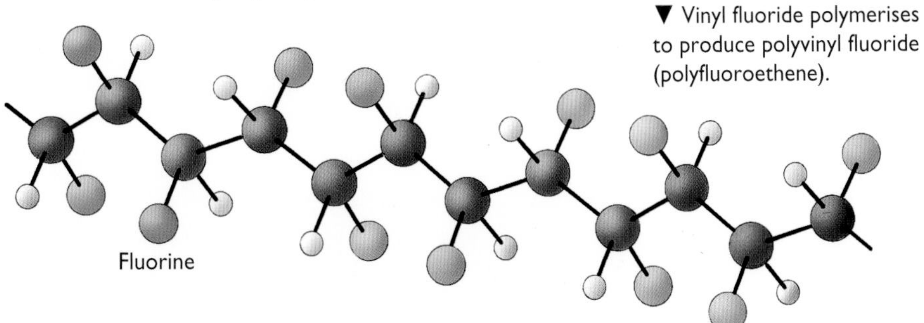

▼ Vinyl fluoride polymerises to produce polyvinyl fluoride (polyfluoroethene).

Chlorates

When chlorine and a large amount of oxygen combine with a metal, the resulting substance is known as a chlorate. Compounds that include chlorates are useful because they decompose to release oxygen.

Potassium chlorate is used as the oxygen supply in fireworks and safety matches. Ammonium perchlorate is used in the booster rockets of the space shuttle.

Sodium and potassium chlorates are used as weedkillers. Sodium and calcium chlorates are also used as oxidising bleaches (see page 14).

Dangers of using chlorate compounds

The large amount of oxygen in chlorate compounds means that they can be potentially very inflammable, and in some cases even explosive.

Weedkillers containing chlorates will, for example, remain as a coating on the weeds they have killed. This is especially dangerous in dry climates where there is always a risk of fire. It is also why some illegal explosives have been made using weedkillers.

▶ Potassium produces a characteristic lilac-coloured flame, showing that it has been used in this catherine wheel firework.

Potassium chlorate and fireworks

Potassium chlorate is an excellent source of the oxygen needed for combustion in fireworks. In addition, by changing the metal ions in the chlorate, the colours characteristic of each metal can be produced, thus providing colour to the display. Strontium chlorate gives a red colour; copper chlorate yields blue; barium chlorate green; and sodium chlorate yellow.

Making sodium chlorate

Sodium chlorate can be produced by the electrolysis of brine, a process similar to that taking place in the diaphragm cell. A variety of chlorates with differing amounts of oxygen can be made.

Sodium chlorate containing the least oxygen is used as a bleach (see page 14). Sodium chlorates with more oxygen, can be used in explosives, matches, fireworks, weedkillers, making oxygen (as shown below) and for throat lozenges.

decompose: to break down a substance (for example by heat or with the aid of a catalyst) into simpler components. In such a chemical reaction only one substance is involved.

explosive: a substance which, when a shock is applied to it, decomposes very rapidly, releasing a very large amount of heat and creating a large volume of gases as a shock wave.

EQUATION: Making sodium chlorate

Sodium chloride + water ⇨ sodium hydroxide + chlorine + hydrogen

$$2NaCl(aq) + 2H_2O(l) \Rightarrow 2NaOH(aq) + Cl_2(g) + H_2(g)$$

Chlorine + sodium hydroxide ⇨ sodium chlorate + water + sodium chloride

$$3Cl_2(g) + 6NaOH(aq) \Rightarrow NaClO_3(aq) + 3H_2O(l) + 5NaCl(aq)$$

Matches

Matches consist of an incendiary (fire-making) device and a source of fuel (the wood of the match stick).

The match head contains potassium chlorate (a source of oxygen). When the match is struck, the head scrapes across the phosphorus on the match box side, causing a small amount of red phosphorus to heat up and come into contact with a source of oxygen. It then reacts and ignites, which in turn causes the match head to begin to burn for long enough to set the wood of the stick on fire.

Preparing oxygen using potassium chlorate

Oxygen is conveniently produced in the laboratory by heating potassium chlorate with a manganese dioxide catalyst.

EQUATION: Using potassium chlorate to make oxygen

Potassium chlorate ⇨ potassium chloride + oxygen

$$2KClO_3(s) \Rightarrow 2KCl(s) + 3O_2(g)$$

▶ This "strike-anywhere" match contains potassium chlorate ($KClO_3$) and tetraphosphorus trisulphide (P_4S_3) so that it does not need a phosphorus strike to be ignited. Potassium chlorate is used rather than the less expensive sodium chlorate because it absorbs less water and is therefore more likely to work when struck.

Chlorine and poison gases

A poison gas is any gaseous chemical that can harm humans. In World War I the Germans made a poison gas attack at Ypres in Belgium. They put chlorine gas in cylinders under pressure and then released it when the wind was blowing towards the trenches containing Allied troops. At this time gas masks had not been invented.

Chlorine affects the cells of all the throat, mouth and lungs, causing damage to all the tissues and also causing fluids to be drawn out of the body and into the lungs. Death follows exposure for more than a couple of minutes.

Even more deadly is a combination of chlorine and cyanide. When this gas is released and inhaled it prevents the red blood cells from carrying away waste carbon dioxide. This causes death within a very short time. Even in small doses, this terrible gas can affect the lungs and cause pneumonia. Mustard gas and phosgene – two of the most feared gases of all – are also based on chlorine compounds.

Tear gas

Less harmful gases that irritate the throat and lungs and cause the tear glands to release fluids are called tear gases. Tear gases are used for crowd control in situations where there is risk to life and property. Many tear gases contain chlorine. One of the most common is CN gas (chloroacetophenone). CS gas (chlorobenzylidenemalononitrile) acts faster and lasts longer than CN gas, causing involuntary closing of the eyes and a burning sensation. However, the effects disappear within a few minutes of being exposed to clean air.

Tear gases are effective at extremely low concentrations (refer back to the sensitivity of people to chlorine, page 10).

▼ Clouds of mustard gas drift across the battlefields of northern France in World War I. In this attack on the 1st July 1916, 94,000 soldiers were killed.

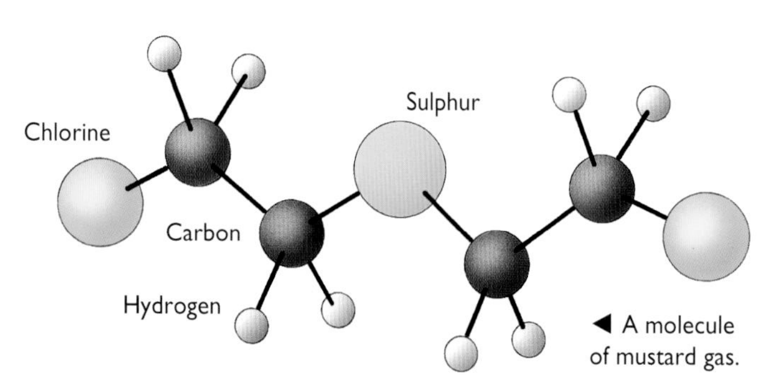

◄ A molecule of mustard gas.

Mustard gas

Mustard gas is 2,2'-dichlorodiethyl sulphide. It is toxic when it falls on the skin and eyes and when breathed into the lungs. The vapour is effective at low concentrations, and because the liquid evaporates slowly, its effects can last in a target area for several weeks.

▲ Gas masks contain filter pads of activated carbon. These absorb the tear or poison gas.

adsorb/adsorption: to "collect" gas molecules or other particles on to the *surface* of a substance. They are not chemically combined and can be removed. (The process is called "adsorption".) Compare to absorb.

toxic: poisonous enough to cause death.

vapour: the gaseous form of a substance that is normally a liquid. For example, water vapour is the gaseous form of liquid water.

Activated charcoal in gas masks

Activated charcoal has a very large surface area (about 2000 sq m of surface area for every gram in weight of charcoal). It is able to take up (adsorb) large numbers of molecules of gases onto this vast surface.

This impressive property has meant that activated charcoal is widely used as gas filters and gas masks in war.

The sequence of pictures below shows a gas jar with activated charcoal in the bottom which has been filled with bromine, a poisonous brown halogen gas. The pictures were taken over a few minutes. The colour indicates the amount of free bromine in the gas jar. Notice that in the gas jar on the far right, there is no free bromine left at all.

Once all the sites on the activated charcoal have been used, it has to be thrown away. It cannot be reactivated.

▶ Pieces of activated charcoal have been dropped into a gas jar containing bromine and the cover glass replaced.

This sequence of photographs shows that at the end of a few minutes the gas jar is colourless, because all of the bromine molecules are now adsorbed on to the surface of the activated charcoal.

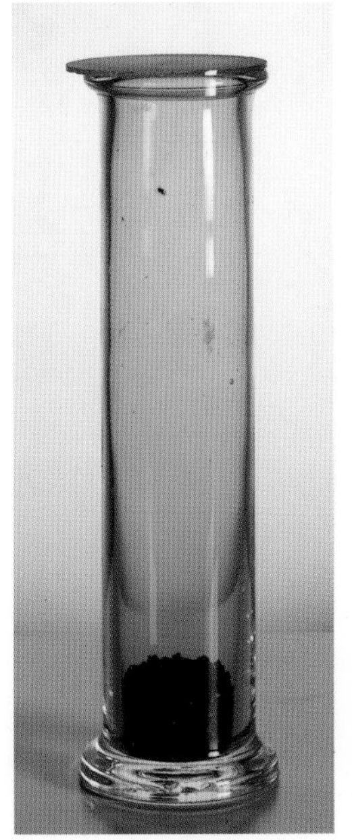

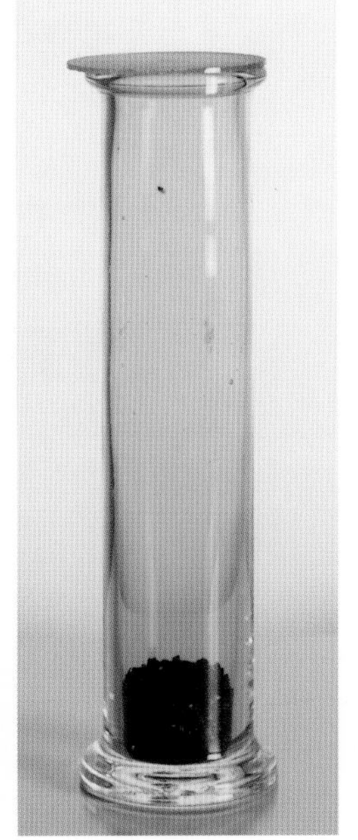

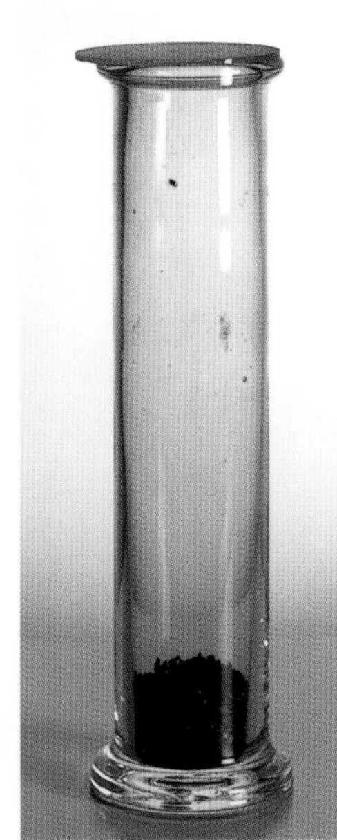

Chlorine and the environment

One of the main uses of chlorine is to combine it with organic substances. This produces organochloride compounds, liquids which can be important solvents (for dissolving other substances), or added to textiles to make them fire resistant. They are also vital as pesticides and fungicides, and they are widely used throughout the world.

The use of these chemicals has made it possible for some pests and diseases to be kept under control. However, most of these substances are extremely poisonous and do not readily break down in the environment. Halogens therefore pose some quite severe environmental problems if used unwisely.

▼ New insecticides are being developed all the time so that they can be more effective and less harmful to the environment.

PCBs

One group of the most notorious organochlorides is known as PCBs (polychlorinated biphenyls). They have been widely used in making electrical appliances, power station transformers, etc. At the time of their use, until a series of accidents occured, nobody realised that waste water containing traces of PCBs was highly dangerous. Much of this material becomes mixed up in the river muds. In one site on the Hudson River in eastern USA, PCBs gradually worked their way up the food chain from worms in the river muds, through fish, to humans until traces of PCB could be found in mother's milk.

The problem exemplified here, however, applies to many chlorine-containing compounds because chlorine reacts with all other materials. No one knows what the effects are of some of the compounds created as a result of something so well meaning as the chlorination in public water supplies.

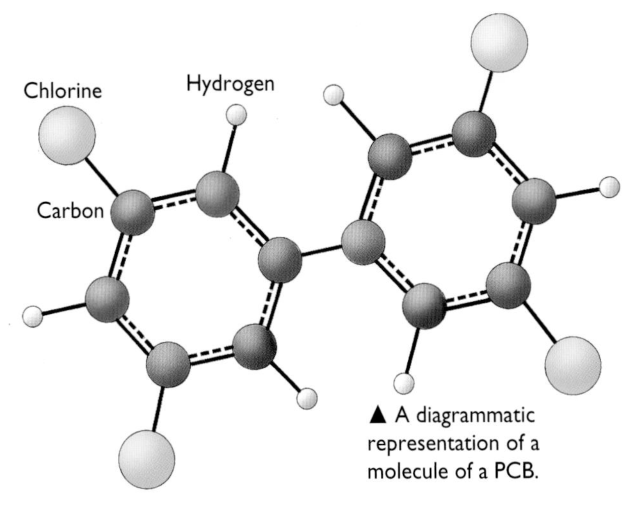

▲ A diagrammatic representation of a molecule of a PCB.

PBBs

Another material, this time a compound of bromine (see page 42), polybrominated biphenyl (PBB), is used to prevent materials catching fire. By accident, in 1973, some of it got mixed in with cattle feed in Michigan, USA, and gradually worked its way up the food chain. Many animals had to be put down, and many people became ill.

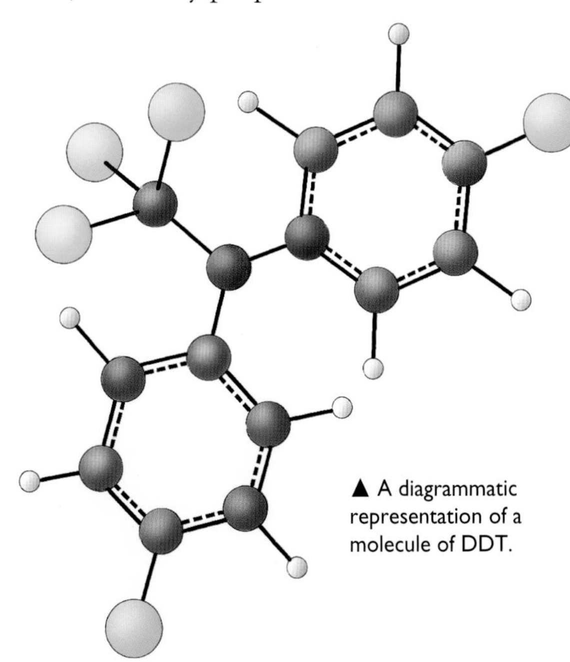
▲ A diagrammatic representation of a molecule of DDT.

DDT

DDT (dichlorodiphenyltrichloroethane) is an insecticide based on chlorine that for many years was used to combat mosquitos. Its use has been stopped for fear of its environmental side effects.

fungicide: any chemical that is designed to kill fungi and control the spread of fungal spores.

pesticide: any chemical that is designed to control pests (unwanted organisms) that are harmful to plants or animals.

Dioxins

Agent Orange is famous for its use in the Vietnam war. It was sprayed onto Vietnamese rainforests because it quickly caused tree leaves to shrivel and die. By this action the American forces hoped to prevent the Viet Cong guerrillas from hiding in the rainforests. Agent Orange is a compound based on chlorine (its chemical description is 2,4,5–trichlorophenoxyethanoic acid and esters).

This chemical contained certain impurities, including another compound of chlorine known as dioxin (2,3,7,8–tetrachlorodibenzo-*p*-dioxin). Dioxin is extremely poisonous, insoluble and clings to the soil, making land unusable for decades.

After Agent Orange was applied in Vietnam, many deformed children were born in the areas that were sprayed. This shows how the dioxin compounds remain active for long periods and find their way up the food chain.

The largest unintentional release of dioxins was in Seveso, Italy, in 1976. Then an escape of dioxins spread out over the area near a chemical plant.

Dioxins were also found in the soil beneath houses in the township of Love Canal, near Niagara Falls, USA. The houses were declared uninhabitable and the citizens forced to move. Dioxins have now been banned.

Paraquat

Paraquat is an organochloride designed as a weedkiller and is used on many household lawns. However, when it was used on a vast scale to kill weeds in the rubber estates of Thailand (4 million kilogrammes were used!), the chemical got into the river water and killed millions of fish. Not only was this an ecological disaster, but fish are a main source of protein for many rural Thais and so the chemical was passed through the food chain to them, causing pesticide poisoning.

Chlorine compounds as solvents

A solvent is a liquid chemical substance that will dissolve another solid substance and yet not react with it. Solvents can also be used to extract one material from another.

Chlorine (and the other halogens), especially when combined with hydrocarbons, easily dissolves organic materials (grease, oil, fat etc.). This makes them very useful solvents for organic materials, because water, the most common solvent, will not dissolve organic materials.

One of the most widely used chlorinated hydrocarbons is trichloroethene. It is used in "dry cleaning" (that is, cleaning without water and using some other liquid solvent instead).

The advantage of trichloroethene over other petroleum-based solvents such as turpentine is that it will not catch fire.

In the dry cleaning process, the solvent is mixed with a detergent. At the end of the cleaning cycle the solvent is recovered from the machine for reuse.

Degreasing agents

Solvents are widely used in engineering factories. Many machines use grease and oil as both a lubricant and coolant. As a result, components often have a thin film of oil or grease on them. This must be removed before they can be coated with paint or other finish, a process called degreasing.

Trichloroethene is the most widely used degreasing agent, often in combination with a detergent.

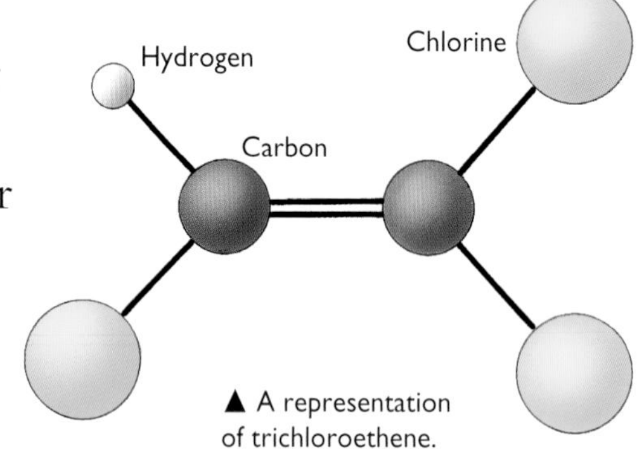

▲ A representation of trichloroethene.

Carbon tetrachloride

Carbon tetrachloride, like trichloroethene, has been widely used as a solvent for removing grease from fabrics. For many years carbon tetrachloride was available for use in the home – often in the form of a jar with a fabric applicator – as a "spot remover".

It has subsequently been discovered that the fumes from carbon tetrachloride are a health hazard, and its sale is now restricted.

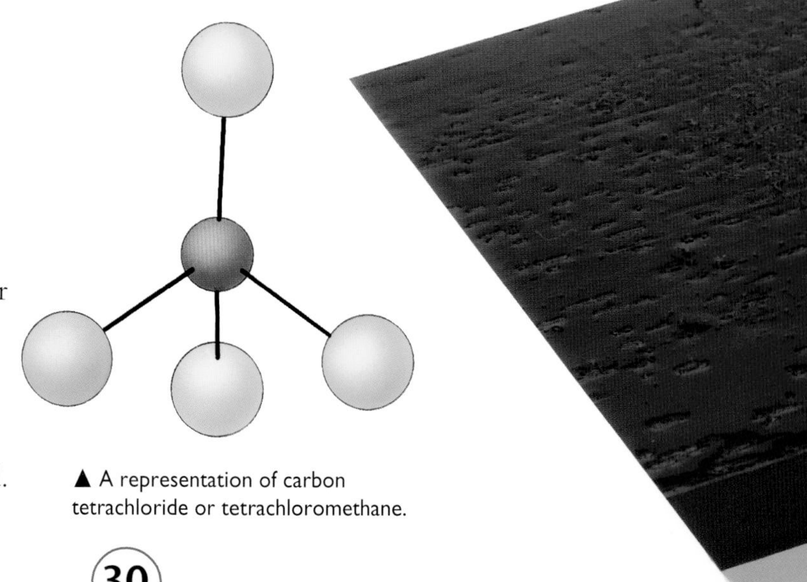

▲ A representation of carbon tetrachloride or tetrachloromethane.

Paint stripping agents

Paint strippers are designed to dissolve paints that are not water soluble, particularly gloss paints.

The most commonly used solvent for this purpose is dichloromethane. Commercial paint strippers mainly use a mixture containing dichloromethane and a gelling agent (whose function is to keep the dichloromethane in contact with paint long enough for the solvent action to work). It is applied to the surface of the paint, which is first softened, then dissolved.

The solvent may require many hours before it completely dissolves the paint, especially in cold conditions. Then the dissolved paint can simply be wiped off. Because the solvent is in contact with the outside of the paint only, premature removal of the solvent leaves behind undissolved paint.

detergent: a petroleum-based chemical that removes dirt.

gelling agent: a semi-solid jelly-like substance.

hydrocarbon: a compound in which only hydrogen and carbon atoms are present. Most fuels are hydrocarbons, as is the simple plastic polyethene (known as polythene).

solvent: the main substance in a solution (e.g. water in salt water).

▶ Paint stripper being used to remove old paint.

◀ A representation of dichloromethane.

Hydrogen chloride gas

Hydrogen chloride gas is produced by reacting the gases hydrogen and chlorine. The reaction is an example of combustion without oxygen.

Hydrogen chloride gas is colourless. The white fumes seen in these pictures are produced as the gas contacts moist air and forms small droplets of hydrochloric acid.

❶▼ Hydrogen gas is produced by reacting an acid on a metal. The small flame shows that the gas is inflammable hydrogen.

EQUATION: Laboratory production of hydrogen

Hydrochloric acid + zinc ➪ zinc chloride + hydrogen

$2HCl(aq)$ + $Zn(s)$ ➪ $ZnCl_2(aq)$ + $H_2(g)$

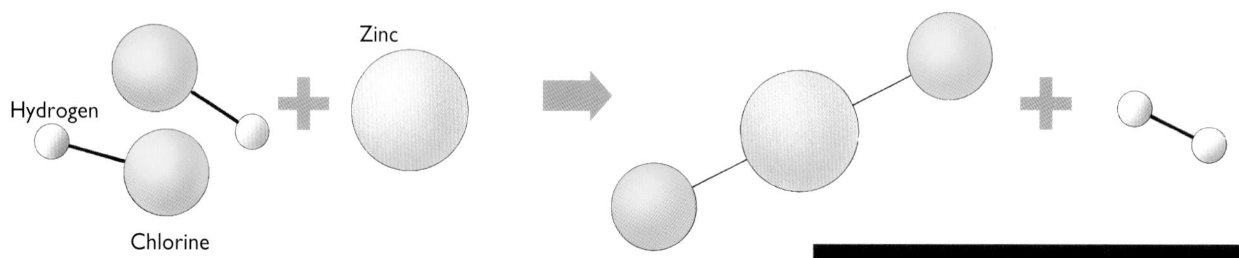

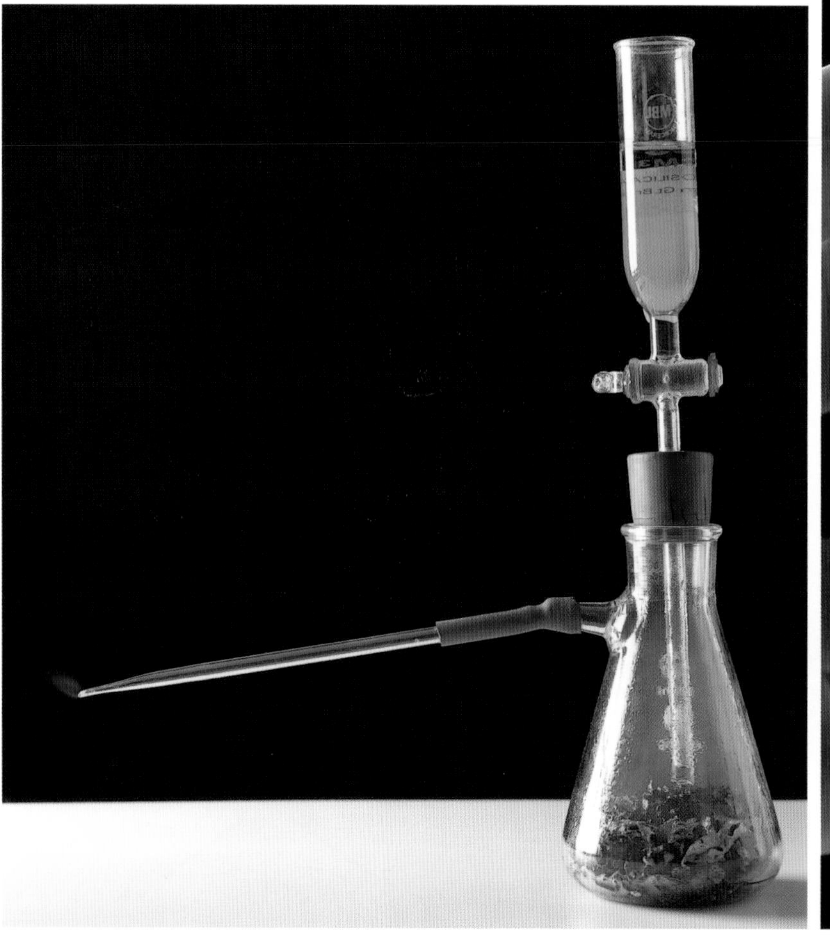

combustion: the special case of oxidisation of a substance where a considerable amount of heat and usually light are given out. Combustion is often referred to as "burning".

EQUATION: Producing hydrogen chloride gas

Hydrogen+ chlorine ➪ hydrogen chloride gas

$H_2(g)$ + $Cl_2(g)$ ➪ $2HCl(g)$

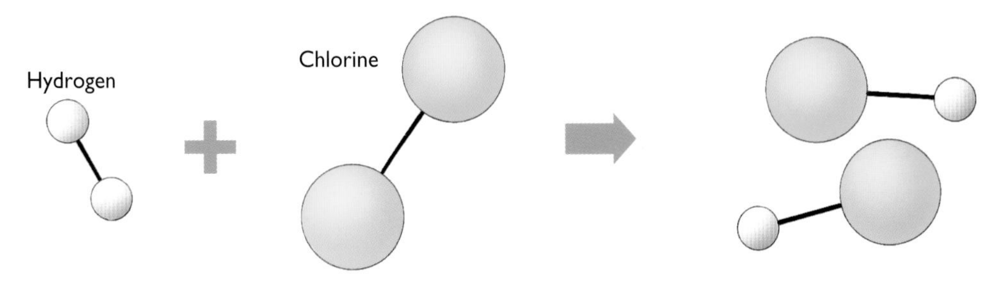

❷▼ Chlorine gas is produced by the method shown on pages 10 and 11. Hydrogen gas (ignited) is introduced into a flask of chlorine gas. The flame becomes white. Notice also a white "mist" at the end of the tube. This is where the hydrogen chloride attracts moisture.

The hydrogen and chlorine react to produce hydrogen chloride gas, which then attracts water and forms a mist of acidic water droplets (hydrochloric acid).

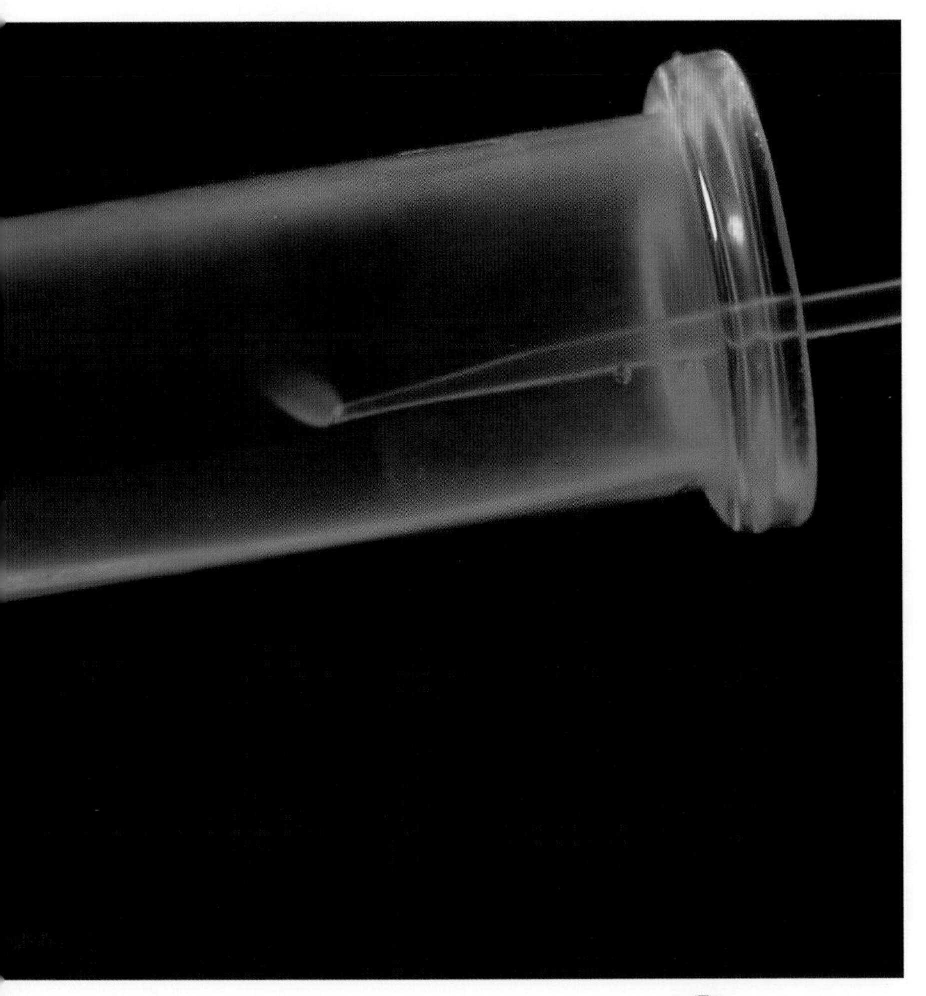

Properties

Hydrogen chloride has a pungent smell. If any of the gas is inhaled it produces immediate, severe irritation of the nose and throat because it dissolves in the mucus membranes and immediately turns into hydrochloric acid.

In addition to the method of preparation shown in the pictures on these pages, hydrogen chloride can be made in the laboratory by reacting concentrated sulphuric acid on sodium chloride (see the equation on page 35).

Vinyl chloride is manufactured from acetylene and hydrogen chloride. When polyvinyl chloride is heated above 150°C, it decomposes, releasing about half its weight as hydrogen chloride gas.

Hydrochloric acid

Hydrochloric acid, previously also called "muriatic acid", is the gas hydrogen chloride (HCl) in water. Hydrochloric acid is a natural, strong mineral acid, found close to the site of erupting volcanoes. It is also one of the main acids involved in human digestion, breaking down the food in our stomachs.

Because hydrochloric acid is a strong acid it is a good electrolyte.

Most hydrochloric acid is produced as a byproduct during the manufacture of plastics. Hydrochloric acid is vital for the manufacture of a common plastic, vinyl (see page 22). It is mainly used in steelworks for cleaning steel, and in extracting metals from their ores.

Demonstration of the reaction of two gases to make a salt

Two colourless gases, one in each cylinder, are brought together. The upper one is ammonia, the lower one hydrogen chloride. Ammonia is a base, so the gases react to form a salt.

The reaction produces a "smoke" of white particles of ammonium chloride (the salt).

❶

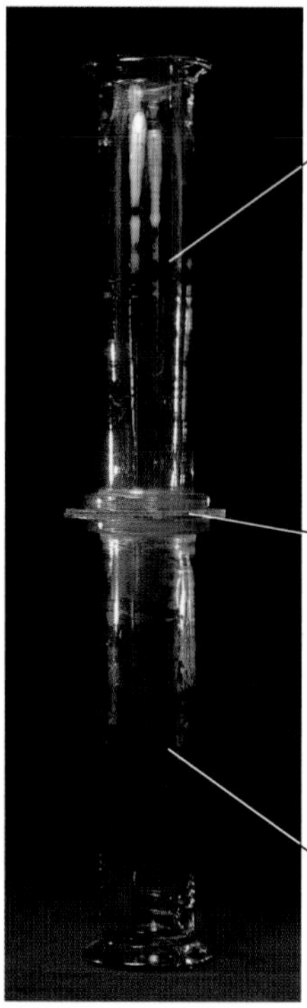

Hydrogen chloride gas

Gas jars placed on top of one another and a blocking glass plate in between. The glass plate is removed to allow the gases to mix and react.

Ammonia gas

❷

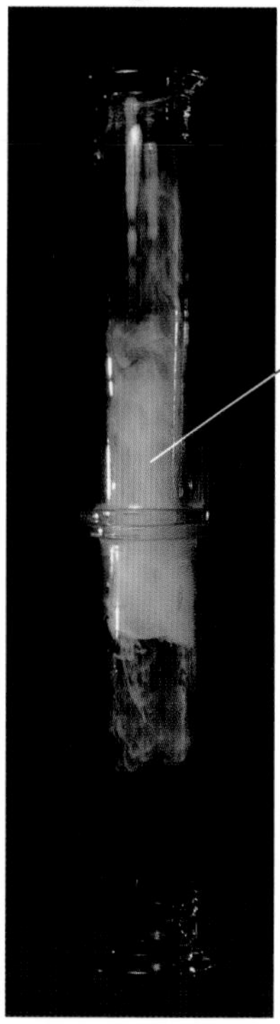

Ammonium chloride smoke appears where the hydrogen chloride and ammonia gas mix and react.

❸

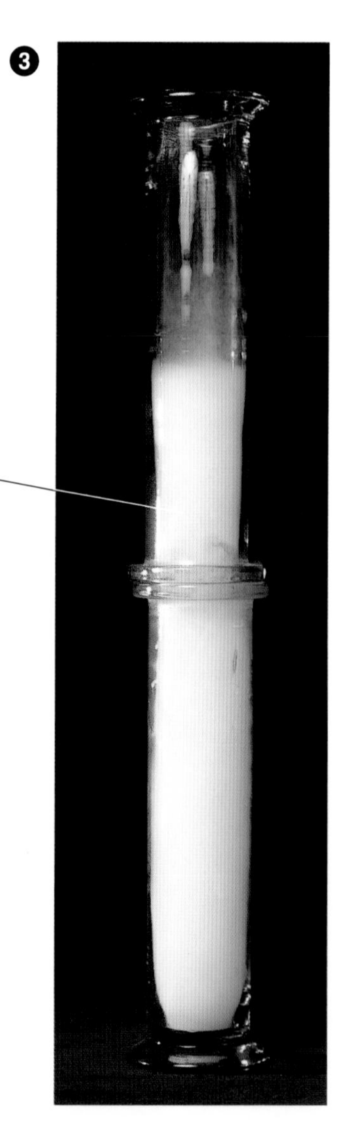

EQUATION: Hydrogen chloride and ammonia

Hydrogen chloride gas+ ammonia gas ⇨ ammonium chloride (solid)

$HCl(g)$ + $NH_3(g)$ ⇨ $NH_4Cl(s)$

base: a compound that may be soapy to the touch and that can react with an acid in water to form a salt and water.

salts: compounds, often involving a metal, that are the reaction products of acids and bases. (Note "salt" is also the common word for sodium chloride, common salt or table salt.)

◀ This picture shows dilute hydrochloric acid being poured onto limestone rock.

These are bubbles of carbon dioxide gas.

EQUATION: Hydrochloric acid and limestone

Hydrochloric acid + calcium carbonate ⇨ calcium chloride + water + carbon dioxide gas

$2HCl(aq) + CaCO_3(s) \Rightarrow CaCl_2(aq) + H_2O(l) + CO_2(g)$

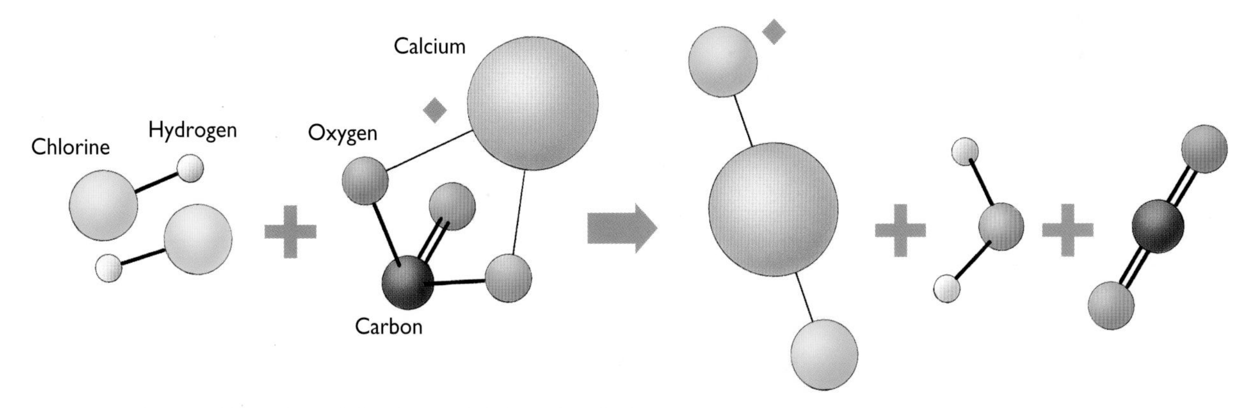

Preparation of hydrochloric acid in the laboratory

Hydrochloric acid is conveniently prepared by reacting concentrated sulphuric acid with sodium chloride. Sodium hydrogen sulphate is precipitated and hydrochloric acid (hydrogen chloride) gas is released.

EQUATION: Sulphuric acid and sodium chloride

Sulphuric acid + sodium chloride ⇨ sodium hydrogen sulphate (precipitate) + hydrogen chloride (gas)

$H_2SO_4(l) + NaCl(s) \Rightarrow NaHSO_4(s) + HCl(g)$

Fluorine

Fluorine, the 13th most abundant element, gets its name from the Latin *fluo*, meaning to flow. This is because, until recent times, fluorine compounds were used as a flux, a material that helps molten metal flow.

Compounds containing fluorine are found in all living things, from plant tissues to animal bones and blood. They are also widely found in water and in rocks. Fluorspar, often called Blue John, is the most common compound of fluorine and contains beautiful crystals (page 8).

Fluorine is a tiny atom and the most reactive of all the elements in the universe, which is why it never occurs in nature on its own. It can only be separated from compounds using large amounts of electrical energy. The fluorine thus produced is a pale yellow gas, highly poisonous and so corrosive that it is extremely difficult to store. Usually a vessel containing fluorine has to be coated inside with a plastic itself made with fluorine.

Compounds of fluorine are called fluorides. Most fluorides do not dissolve in water or oil and they do not catch fire. Uranium hexafluoride was used in making the first atomic bomb.

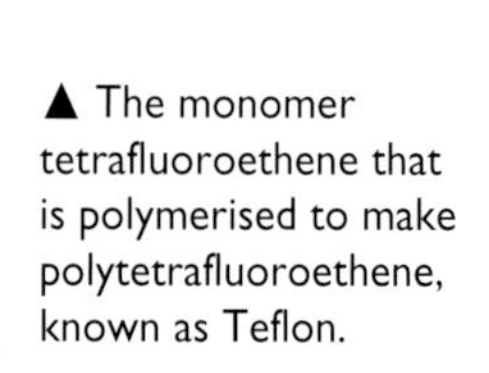

▲ The monomer tetrafluoroethene that is polymerised to make polytetrafluoroethene, known as Teflon.

Tetrafluoroethene

Tetrafluoroethene is a material containing fluorine that has proved very popular. It is a polymer made from an extract of crude oil (ethene) and fluorine.

Polytetrafluoroethene is a soft white plastic (polymer) that virtually nothing sticks to. It is nonreactive with most chemicals (that is, it resists corrosion), heat resistant, and a very good electrical insulator. It is the nonstick surface used on many household cooking utensils and cookware.

Because of all these useful properties, tetrafluoroethene is used worldwide for tubes, gaskets and seals that must not deteriorate. It is also used for making bearings and as a coating on saw blades.

This material is one of the main sources of demand for fluorine.

▲ Polytetrafluoroethene coatings are found on many household pans to give them their nonstick finish.

The only problem with the material is that it is soft and so can be scratched with metal implements. For this reason wooden or plastic implements need to be used with nonstick pans.

corrosive: a substance, either an acid or an alkali, that *rapidly* attacks a wide range of other substances.

flux: a material used to make it easier for a liquid to flow. A flux dissolves metal oxides and so prevents a metal from oxidising while being heated.

polymer: a compound that is made of long chains by combining molecules (called monomers) as repeating units. ("Poly" means many, "mer" means part).

resin: natural or synthetic polymers that can be moulded into solid objects or spun into thread.

Fluoride and teeth

One of the most remarkable uses of a chemical has been the widespread use of compounds of fluorine in drinking water and toothpaste.

Teeth have a natural coating called enamel, made of calcium phosphate. This is quite a durable substance; however, tooth decay from eating too much sugar has become a serious problem in many countries.

Fluorides occur naturally in many water supplies, and scientists in the United States were able to show that in areas where there was more fluoride in the natural drinking water, there was a lower incidence of tooth decay. This is because fluoride strengthens the tooth enamel, creating a complex of calcium, fluoride and phosphate ions.

Fluoride is now added to many water supplies. However, too high a level of fluoride in the water may cause teeth to become very hard and brittle and to develop a dark brown colour. As a result it is important to control the level to about one part per million.

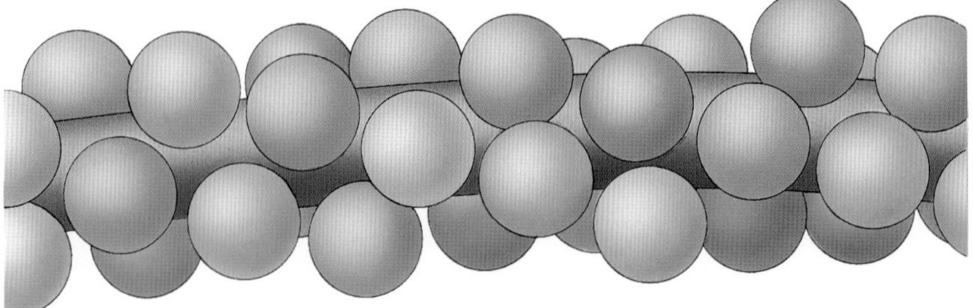

▲ Chains of units like the one shown above form solid polytetrafluoroethene. The fluorine atoms shield the carbon atoms that make the chain (polymer). Thus the carbon atoms form a polymer that is shielded from reaction by the fluorine: the secret of the material's nonstick success.

Also...

Fluorine is corrosive enough to be one of the few chemicals to react with diamond. In fact, fluorine is so reactive that it can actually extract oxygen from water. If fluorine gas is bubbled through water, the fluorine takes the hydrogen atoms from the water molecules to make hydrogen fluoride, or hydrofluoric acid, the world's most corrosive and dangerous gas. At the same time it releases oxygen as a gas.

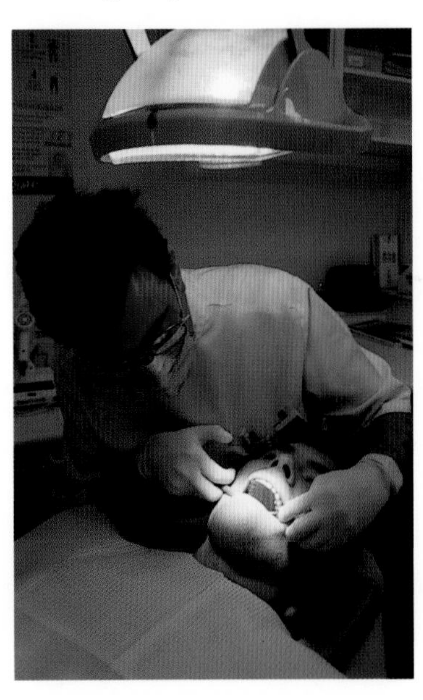

CFCs

Fluorine and chlorine have been widely used to make the coolant in refrigerators. Refrigerants are varieties of substances that include chlorine, fluorine and carbon, called chlorofluorocarbons or more commonly just CFCs.

The properties of CFCs were found to be excellent for use in a number of industries. They are not poisonous, they do not catch fire, they absorb large amounts of heat, and they can be made to change from liquid to gas at temperatures that are just right for use in refrigerators. Because CFCs do not easily react with other chemicals, they also proved to be very good for use in aerosol cans, acting as the gas that expels the liquid from the can in a fine mist.

The first CFCs were made in the 1930s. However, by the 1970s it was becoming clear that CFCs were not as safe as had once been thought. Their continued and escalating use was destroying the ozone layer high in the atmosphere. As a result, CFCs are being replaced with other refrigerants and gases as fast as is practical.

▲ The gradual enlargement of the ozone hole – the region of the upper stratosphere that has been depleted of ozone – is thought to be largely the result of ozone reacting with the CFC molecules released from the ground. Recent attempts to limit the production of CFCs are designed to prevent more destruction of the ozone, a vital shield against the cancer-producing ultraviolet rays of the Sun.

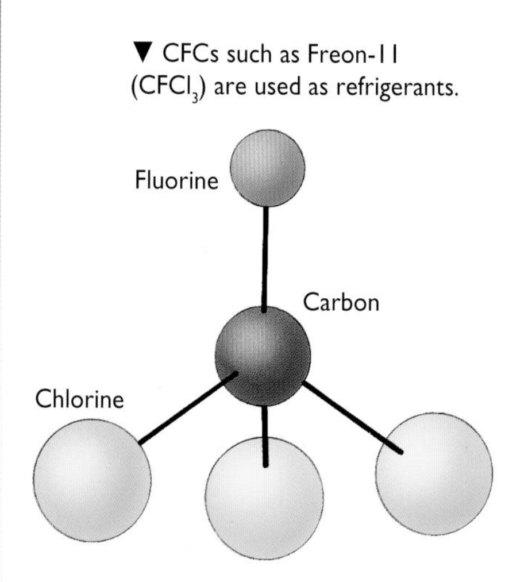

▼ CFCs such as Freon-11 ($CFCl_3$) are used as refrigerants.

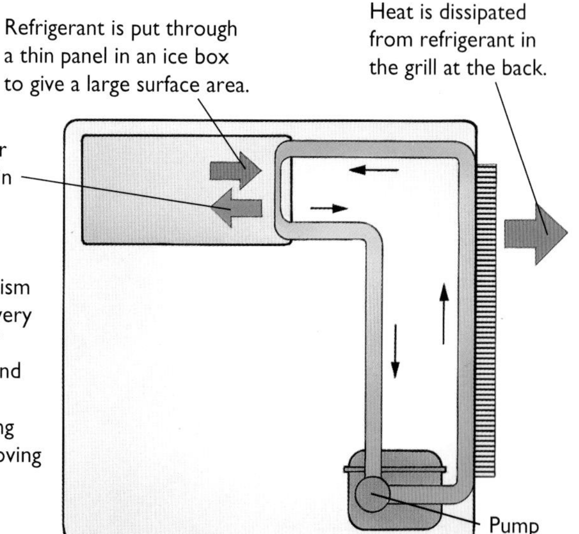

► A diagram of the mechanism of a refrigerator. CFCs are very efficient coolants because of their high thermal capacity and boiling points. They are also cheap to manufacture. Finding replacement chemicals is proving to be extremely difficult.

▼ When chlorofluorocarbons (CFCs) reach the stratosphere they are broken down by ultraviolet solar radiation releasing chlorine atoms. These atoms attack the ozone molecules. A single atom can survive in the stratosphere for four to ten years, and during that time it can destroy countless ozone molecules. It is a three-stage process, as shown by the equations below.

EQUATION: Reaction of CFCs with ozone

❶ *CFC + ultraviolet light ⇨ chlorine*

CFC(g) + UV light ⇨ Cl(g)

❷ *Chlorine atom + ozone ⇨ chlorine oxide + oxygen*

Cl(g) + O_3(g) ⇨ ClO(g) + O_2(g)

❸ *Chlorine oxide + oxygen atom ⇨ chlorine atom + oxygen*

ClO(g) + O(g) ⇨ Cl(g) + O_2(g)

ozone: a form of oxygen whose molecules contain three atoms of oxygen. Ozone is regarded as a beneficial gas when high in the atmosphere because it blocks ultraviolet rays. It is a harmful gas when breathed in, so low level ozone, which is produced as part of city smog, is regarded as a form of pollution. The ozone layer is the uppermost part of the stratosphere.

stratosphere: the part of the Earth's atmosphere that lies immediately above the region in which clouds form. It occurs between 12 and 50 km above the Earth's surface.

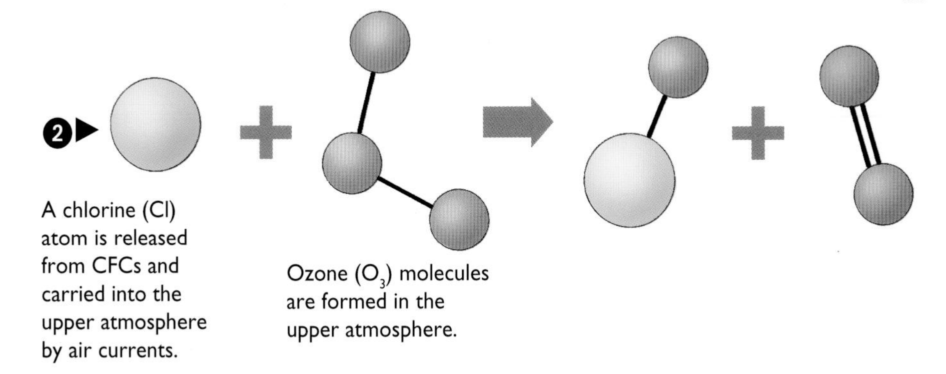

A chlorine (Cl) atom is released from CFCs and carried into the upper atmosphere by air currents.

Ozone (O_3) molecules are formed in the upper atmosphere.

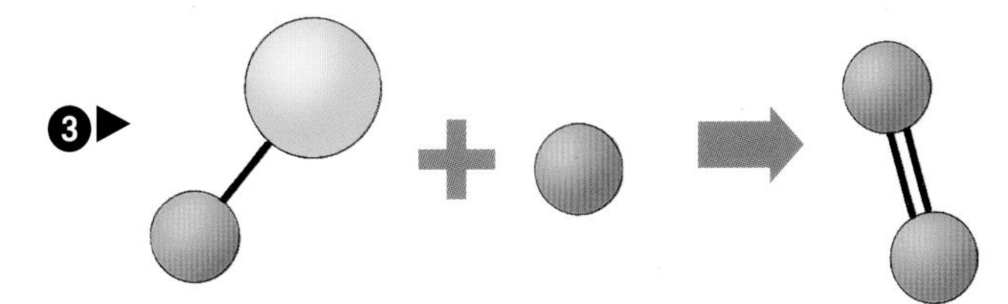

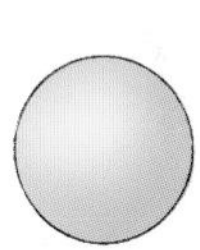

This very reactive chlorine atom is called a "free radical". It can now react with another ozone molecule, causing further destruction of the ozone layer.

How refrigerants work

A refrigerator works by taking heat from one part of its surroundings, and giving it out elsewhere. In general, refrigerators take heat from the area inside them, near the ice box, and give it out to the room through a network of pipes fixed to the back of the refrigerator.

To act as a refrigerant, the liquid used has to be pumped around a system of closed tubes. The liquid is made to boil inside the refrigerator and to condense outside.

The liquid chosen has to be able to get very cold inside the refrigerator (so that is can take up heat from the air in the ice compartment), and get very hot outside it (so it can give heat out to the surroundings). Ammonia is one common gas that could be used, but the inert (nonreacting) properties of CFCs are preferable.

▲ The first CFC-free substitutes are now being produced. This one is called KLEA.

Iodine

Iodine is named after the violet colour of its vapour. It is a blue–black solid at room temperature, and is very poorly soluble in water, although its compounds are all soluble. Tincture of iodine, often seen on the bathroom shelf as a dark brown liquid, is iodine dissolved in alcohol. In this form it makes a useful disinfectant.

Iodine is poisonous in high concentrations, but health-giving in trace amounts. It is an essential element to both plants and animals. In animals it is found in the thyroid gland in the throat, where it is used to make vital chemicals that control the rate of body activity. Without sufficient iodine, body activity is slowed down and stunted growth results. Lack of iodine also causes the throat to swell, called a goitre. For these reasons, iodine is added to almost all table salt on sale to the public (sometimes labelled as iodised salt).

Iodine is quite common in the world, especially in sea water. However, it does not make up any kind of rock, so no large mineable deposits exist. It is mainly recovered from ancient salt deposits.

Seaweeds

A number of sea plants contain large natural concentrations of iodine. Sea sponges were traditionally used to reduce the swelling of goitres.

Later it was discovered that the more common seaweed, kelp, also contains large amounts of iodine. Seaweed can be eaten to ensure sufficient iodine in a diet.

Iodine and rain

Iodine combines with many metals, but one of the most important is silver. Silver iodide crystals have been used in rain-making experiments because they attract moisture and act as condensation nuclei – particles that can "grow" the water droplets that fall as rain.

▶ Here you can see the particles of solid iodine in a tube containing water. Clearly, iodine does not dissolve well in water; however, almost all iodine compounds are soluble.

Chlorine displaces iodine

The halogens form a family of elements with similar properties and that form similar compounds. When put together, the more reactive will displace those that are less reactive as shown to the left and below. When chlorine is bubbled through colourless potassium iodide solution, it turns brown. This colour is caused by tiny solid crystals of iodine that will, over time, settle out on the bottom of the beaker, leaving a clear solution of potassium chloride.

condensation nuclei: microscopic particles of dust, salt and other materials suspended in the air, which attract water molecules.

vapour: the gaseous form of a substance that is normally a liquid. For example, water vapour is the gaseous form of liquid water.

▼ Chlorine gas is produced and bubbled through a beaker filled with a solution of potassium iodide. The brown colour is the iodine displaced by the chlorine and released into the solution.

Iodine and health

Iodine has long been used as a mild antiseptic and disinfectant. You will also find iodine combined with some detergents that claim to disinfect as well as clean.

EQUATION: Displacement of iodine by chlorine

Potassium iodide + chlorine gas ➪ potassium chloride + iodine

$2KI(aq) + Cl_2(g) \Rightarrow 2KCl(aq) + I_2(s)$

Also...

Fluorine is the most reactive of the halogens, followed by chlorine, bromine and then iodine as we descend the group in the Periodic Table (see page 47).

Bromine

Bromine, a dark red liquid, is named for the Greek word meaning "stench". Bromine is never found as an element alone because, like the other elements in this book, it is very reactive. Bromine occurs as ions in sea water, and it is normally obtained as a byproduct of refining common salt.

Like iodine, bromine is concentrated in the bodies of some sea life. Both kelp and shellfish can be sources of bromine. The sea mollusc murex was traditionally used to make a purple dye, which was a compound of bromine. People in the Victorian era drank a potion containing a compound of bromine to help them sleep.

Bromine is still used in dyes. It was used in leaded fuel, but this is declining with the use of unleaded fuels. It is perhaps best known for its use as silver bromide in photographic papers.

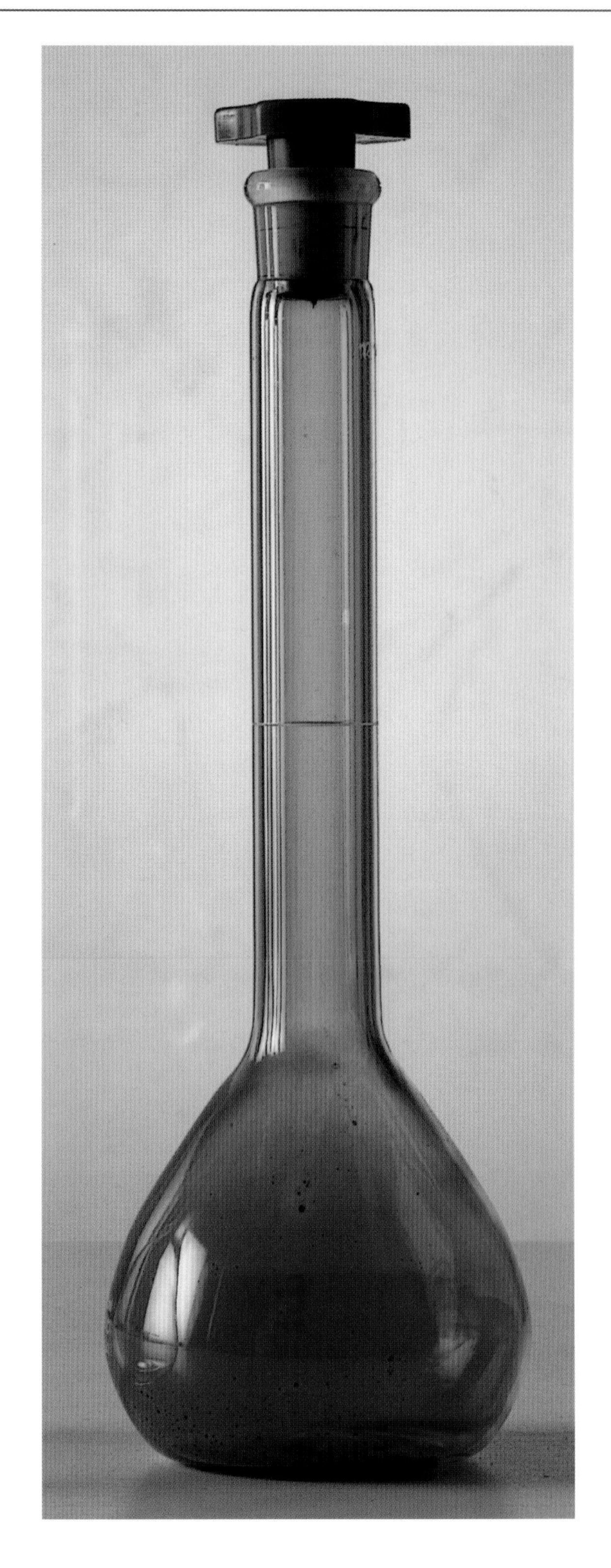

▶ Bromine gas.

Manufacturing bromine

Bromine can be made from sea water, using the property that chlorine will displace bromine in compounds.

About 22,000 tonnes of sea water have to be processed to collect one tonne of bromine.

EQUATION: Obtaining bromine from sea water

Chlorine gas + bromine ions in water ⇨ bromine + chloride ions in water

$Cl_2(g) + 2Br^-(aq) \Rightarrow Br_2(g) + 2Cl^-(aq)$

Bromide in photography

Film for black-and-white photography consists of a transparent plastic sheet on which is spread a thin layer of an emulsion consisting of a suspension of minute mixed crystals of silver bromide and silver iodide in gelatin. The gelatin not only holds the grains but also greatly increases their sensitivity to light. When light shines on this material, the silver aggregates into visible particles. These will form the image. Black and white processing paper also contains silver bromide.

emulsion: tiny droplets of one substance dispersed in another. A common oil in water emulsion is milk. The tiny droplets in an emulsion tend to come together, so another stabilising substance is often needed to wrap the particles of grease and oil in a stable coat. Soaps and detergents are such agents. Photographic film is an example of a solid emulsion.

EQUATION: Silver bromide and photographers' hypo

Sodium thiosulphate (photographers' hypo) + silver bromide ⇨ silver complex + sodium bromide

$2Na_2S_2O_3(aq) + AgBr(s) \Rightarrow Na_3Ag(S_2O_3)_2(aq) + NaBr(aq)$

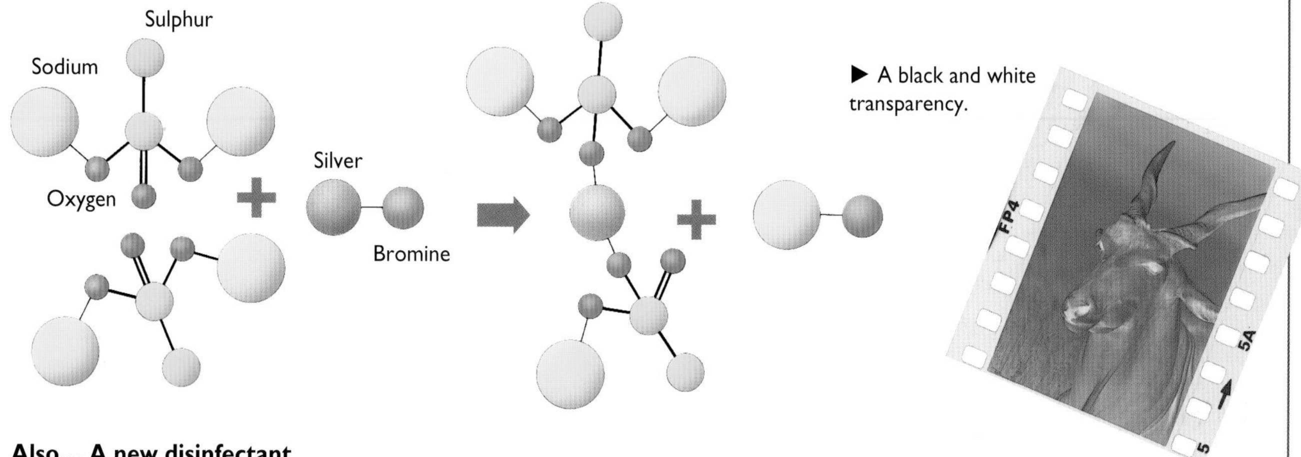

► A black and white transparency.

Also... A new disinfectant

Bromine is part of a new disinfectant for swimming pools that has been designed to get rid of several problems caused by the use of chlorine alone. Chlorine is only effective for about a day, and it is irritating to the eyes because it has to be used in quite large concentrations.

The new disinfectant (1-bromo-3-chloro-4,4,5,5,-tetramethyl-2-imidazolidinine, or BCI for short) is colourless, has no taste or smell and lasts for about a month between treatments. The reason that chlorine has side effects is that it readily ionises in water and reacts with organic impurities. The methyl groups in the new disinfectant stop the compound from breaking down as quickly.

When bromine does break free, it acts as a rapid disinfectant, while the rest of the molecule, containing the chlorine, acts as a longer-acting disinfectant.

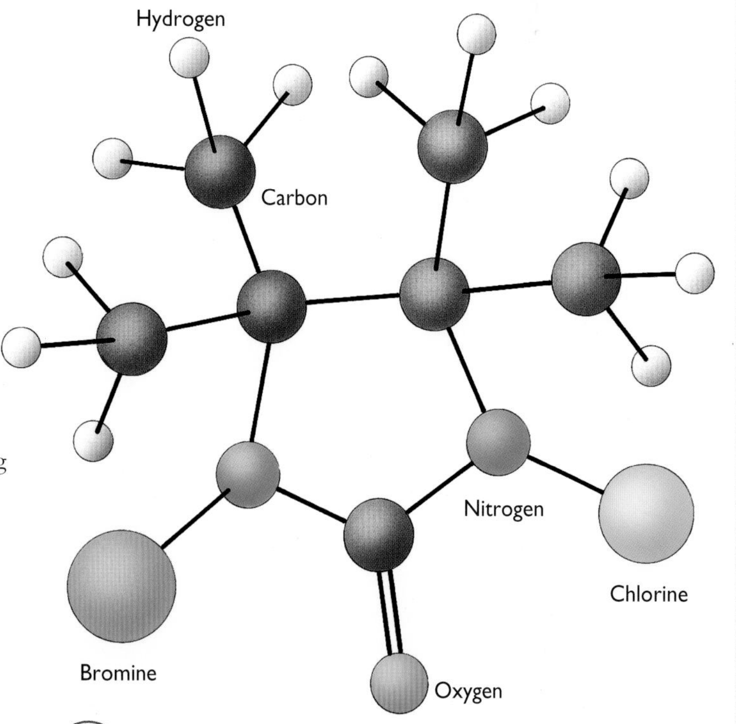

Also...

All of the halogens can be made into useful solvents. Organic bromine complexes are used as paint strippers. Bromine is also used in compounds designed as flame retardants.

Key facts about...

Chlorine

Greenish yellow gas,
chemical symbol Cl

Chloride ions make up
0.015% of the Earth's crust

Pungent smell

Found in common salt

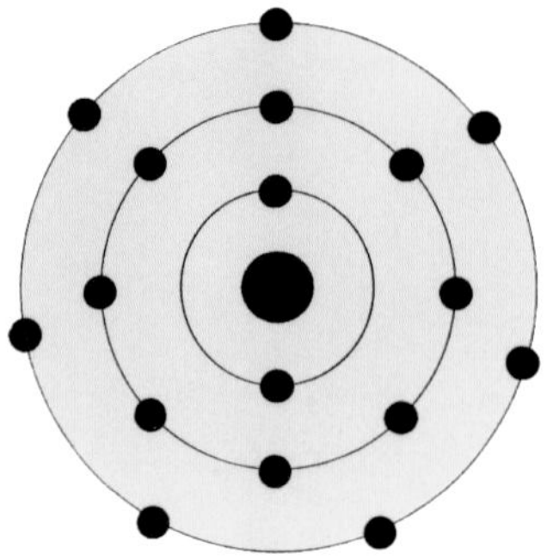

About two and a half
times as dense as air

Slightly soluble
in water

The foundation for
hydrochloric acid

Chloride ions
make up nearly
2% of sea water

Atomic number 17,
atomic weight about 35

Iodine

Violet liquid,
chemical symbol I

Poisonous in large
amounts, but essential
as a trace element to all
plants and animals

17th most abundant
element in sea water;
62nd in the Earth's crust

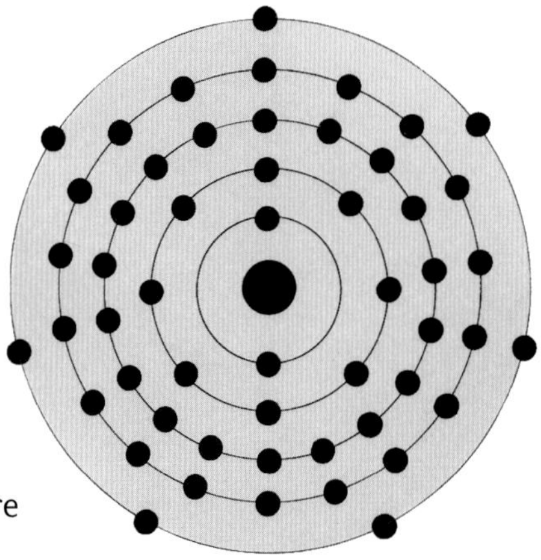

Mainly used as a
disinfectant

A black solid at
room temperature

Atomic number 53,
atomic weight about 127

Bromine

Brown gas, chemical symbol Br

48th most common element on Earth

Powerful unpleasant smell

Mainly used for photography as silver bromide

Makes up 0.0065% of sea water

Bromine compounds were once given as sedatives

Used in flame retardants

Atomic number 35, atomic weight about 80

SHELL DIAGRAMS

The shell diagrams on these pages are representations of an atom of each element. The total number of electrons are shown in the relevant orbitals, or shells, around the central nucleus.

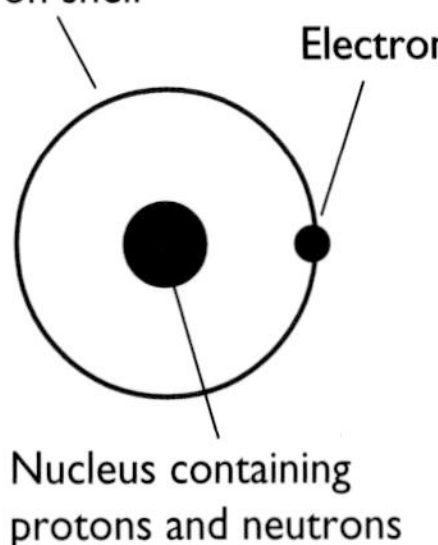

Fluorine

Pale yellow gas, chemical symbol F

The most reactive of all elements

Extremely poisonous in elemental form

Used with carbon-based compounds to make nonstick plastic

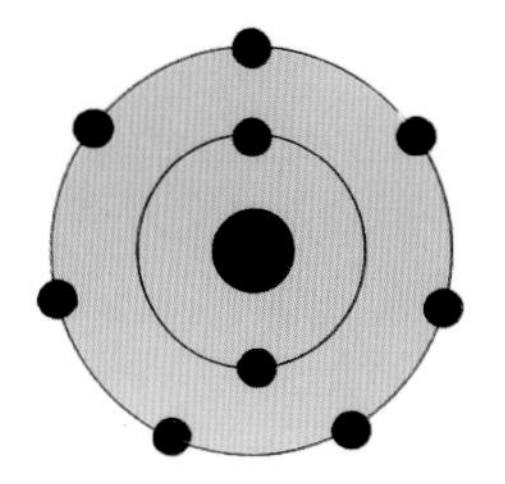

Highly corrosive as a gas and as hydrofluoric acid

Extensively used to prevent tooth decay

Exists freely only as a gas

Atomic number 9, atomic weight about 19

The Periodic Table

The Periodic Table sets out the relationships among the elements of the Universe. According to the Periodic Table, certain elements fall into groups. The pattern of these groups has, in the past, allowed scientists to predict elements that had not at that time been discovered. It can still be used today to predict the properties of unfamiliar elements.

The Periodic Table was first described by a Russian teacher, Dmitry Ivanovich Mendeleev, between 1869 and 1870. He was interested in writing a chemistry textbook, and wanted to show his students that there were certain patterns in the elements that had been discovered. So he set out the elements (of which there were 57 at the time) according to their known properties. On the assumption that there was pattern to the elements, he left blank spaces where elements seemed to be missing. Using this first version of the Periodic Table, he was able to predict in detail the chemical and physical properties of elements that had not yet been discovered. Other scientists began to look for the missing elements, and they soon found them.

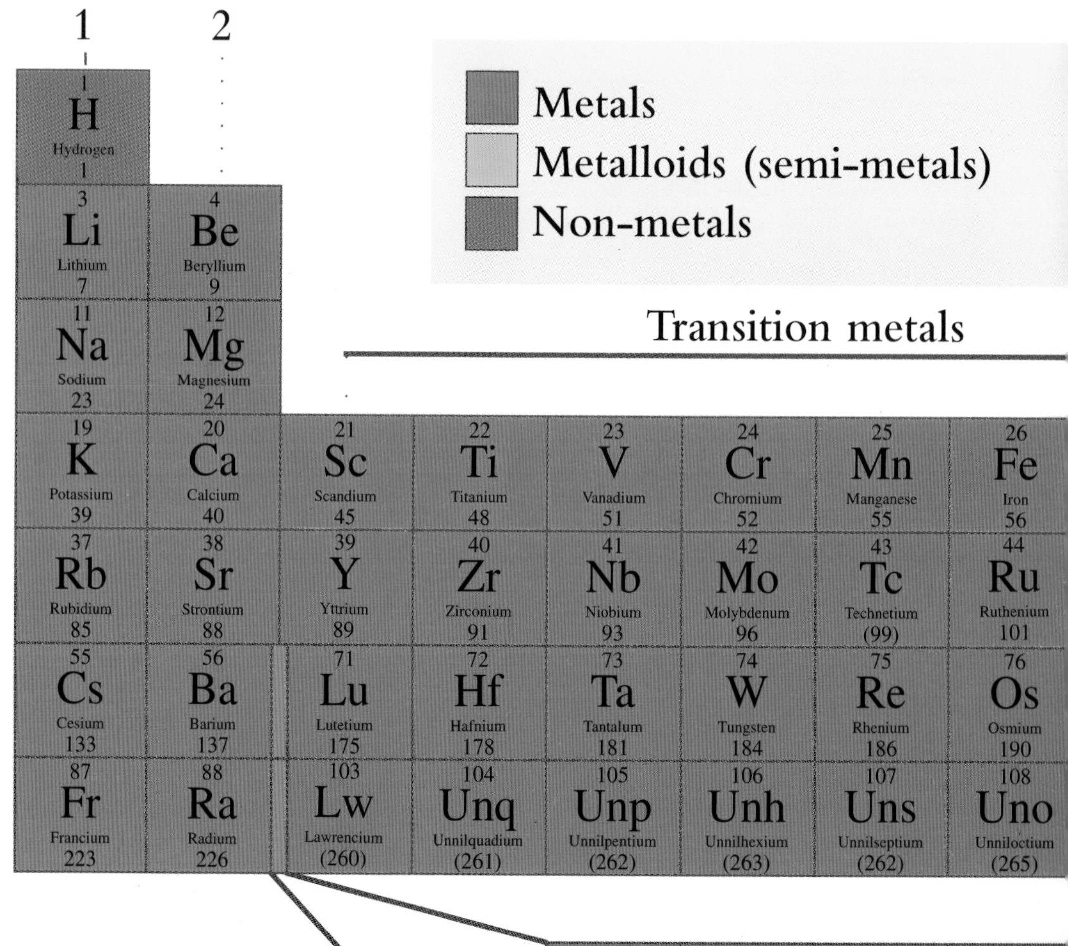

Hydrogen did not seem to fit into the table, so he placed it in a box on its own. Otherwise the elements were all placed horizontally. When an element was reached with properties similar to the first one in the top row, a second row was started. By following this rule, similarities among the elements can be found by reading up and down. By reading across the rows, the elements progressively increase their atomic number. This number indicates the number of positively charged particles (protons) in the nucleus of each atom. This is also the number of negatively charged particles (electrons) in the atom.

The chemical properties of an element depend on the number of electrons in the outermost shell.

Atoms can form compounds by sharing electrons in their outermost shells. This explains why atoms with a full set of electrons (like helium, an inert gas) are unreactive, whereas atoms with an incomplete electron shell (such as chlorine) are very reactive. Elements can also combine by the complete transfer of electrons from metals to non-metals and the compounds formed contain ions.

Radioactive elements lose particles from their nucleus and electrons from their surrounding shells. As a result their atomic number changes and they become new elements.

Atomic (proton) number — 13
Symbol — Al
Name — Aluminium
Approximate relative atomic mass (Approximate atomic weight) — 27

				3	4	5	6	7	0
									2 He Helium 4
				5 B Boron 11	6 C Carbon 12	7 N Nitrogen 14	8 O Oxygen 16	9 F Fluorine 19	10 Ne Neon 20
				13 Al Aluminium 27	14 Si Silicon 28	15 P Phosphorus 31	16 S Sulphur 32	17 Cl Chlorine 35	18 Ar Argon 40
27 Co Cobalt 59	28 Ni Nickel 59	29 Cu Copper 64	30 Zn Zinc 65	31 Ga Gallium 70	32 Ge Germanium 73	33 As Arsenic 75	34 Se Selenium 79	35 Br Bromine 80	36 Kr Krypton 84
45 Rh Rhodium 103	46 Pd Palladium 106	47 Ag Silver 108	48 Cd Cadmium 112	49 In Indium 115	50 Sn Tin 119	51 Sb Antimony 122	52 Te Tellurium 128	53 I Iodine 127	54 Xe Xenon 131
77 Ir Iridium 192	78 Pt Platinum 195	79 Au Gold 197	80 Hg Mercury 201	81 Tl Thallium 204	82 Pb Lead 207	83 Bi Bismuth 209	84 Po Polonium (209)	85 At Astatine (210)	86 Rn Radon (222)
109 Une Unnilennium (266)									

61 Pm Promethium (145)	62 Sm Samarium 150	63 Eu Europium 152	64 Gd Gadolinium 157	65 Tb Terbium 159	66 Dy Dysprosium 163	67 Ho Holmium 165	68 Er Erbium 167	69 Tm Thulium 169	70 Yb Ytterbium 173
93 Np Neptunium (237)	94 Pu Plutonium (244)	95 Am Americium (243)	96 Cm Curium (247)	97 Bk Berkelium (247)	98 Cf Californium (251)	99 Es Einsteinium (252)	100 Fm Fermium (257)	101 Md Mendelevium (258)	102 No Nobelium (259)

Understanding equations

As you read through this book, you will notice that many pages contain equations using symbols. If you are not familiar with these symbols, read this page. Symbols make it easy for chemists to write out the reactions that are occurring in a way that allows a better understanding of the processes involved.

Symbols for the elements

The basis of the modern use of symbols for elements dates back to the 19th century. At this time a shorthand was developed using the first letter of the element wherever possible. Thus "O" stands for oxygen, "H" stands for hydrogen and so on. However, if we were to use only the first letter, then there could be some confusion. For example, nitrogen and nickel would both use the symbols N. To overcome this problem, many elements are symbolised using the first two letters of their full name, and the second letter is lowercase. Thus although nitrogen is N, nickel becomes Ni. Not all symbols come from the English name; many use the Latin name instead. This is why, for example, gold is not G but Au (for the Latin *aurum*) and sodium has the symbol Na, from the Latin *natrium*.

Compounds of elements are made by combining letters. Thus the molecule carbon

Written and symbolic equations

In this book, important chemical equations are briefly stated in words (these are called word equations), and are then shown in their symbolic form along with the states.

What reaction the equation illustrates

Word equation

Symbol equation

EQUATION: The formation of calcium hydroxide

Calcium oxide + water ⇨ calcium hydroxide

$CaO(s) + H_2O(l) \Rightarrow Ca(OH)_2(aq)$

heated

Sometimes you will find additional descriptions below the symbolic equation.

Symbol showing the state: *s* is for solid, *l* is for liquid, *g* is for gas and *aq* is for aqueous.

Diagrams

Some of the equations are shown as graphic representations.

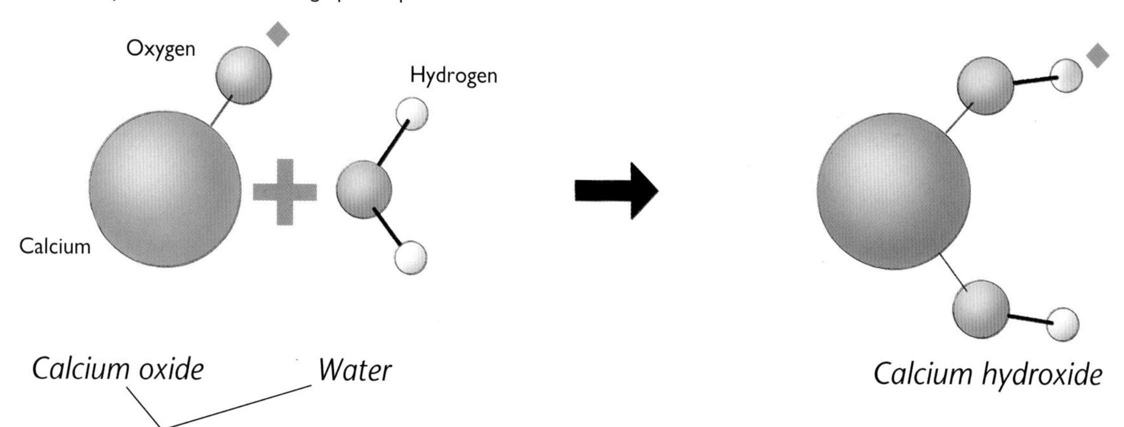

Sometimes the written equation is broken up and put below the relevant stages in the graphic representation.

monoxide is CO. By using lowercase letters for the second letter of an element, it is possible to show that cobalt, symbol Co, is not the same as the molecule carbon monoxide, CO.

However, the letters can be made to do much more than this. In many molecules, atoms combine in unequal numbers. So, for example, carbon dioxide has one atom of carbon for every two of oxygen. This is shown by using the number 2 beside the oxygen, and the symbol becomes CO_2.

In practice, some groups of atoms combine as a unit with other substances. Thus, for example, calcium bicarbonate (one of the compounds used in some antacid pills) is written $Ca(HCO_3)_2$. This shows that the part of the substance inside the brackets reacts as a unit and the "2" outside the brackets shows the presence of two such units.

Some substances attract water molecules to themselves. To show this a dot is used. Thus the blue form of copper sulphate is written $CuSO_4.5H_2O$. In this case five molecules of water attract to one of copper sulphate.

Atoms and ions
Each sphere represents a particle of an element. A particle can be an atom or an ion. Each atom or ion is associated with other atoms or ions through bonds – forces of attraction. The size of the particles and the nature of the bonds can be extremely important in determining the nature of the reaction or the properties of the compound.

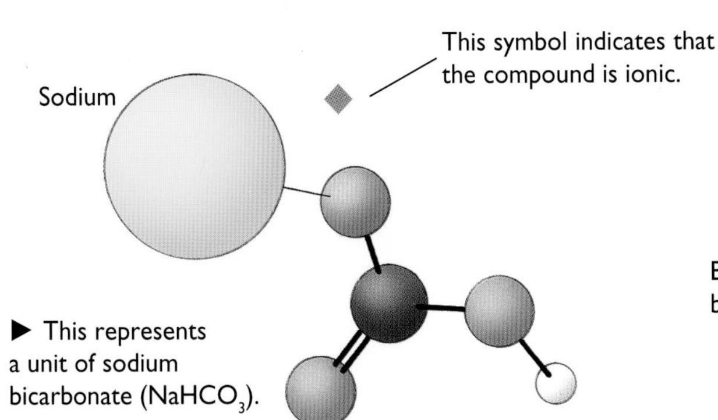

The term "unit" is sometimes used to simplify the representation of a combination of ions.

When you see the dot, you know that this water can be driven off by heating; it is part of the crystal structure.

In a reaction substances change by rearranging the combinations of atoms. The way they change is shown by using the chemical symbols, placing those that will react (the starting materials, or reactants) on the left and the products of the reaction on the right. Between the two, chemists use an arrow to show which way the reaction is occurring.

It is possible to describe a reaction in words. This gives word equations, which are given throughout this book. However, it is easier to understand what is happening by using an equation containing symbols. These are also given in many places. They are not given when the equations are very complex.

In any equation both sides balance; that is, there must be an equal number of like atoms on both sides of the arrow. When you try to write down reactions, you, too, must balance your equation; you cannot have a few atoms left over at the end!

The symbols in brackets are abbreviations for the physical state of each substance taking part, so that (*s*) is used for solid, (*l*) for liquid, (*g*) for gas and (*aq*) for an aqueous solution, that is, a solution of a substance dissolved in water.

Chemical symbols, equations and diagrams
The arrangement of any molecule or compound can be shown in one of the two ways shown below, depending on which gives the clearer picture. The left-hand diagram is called a ball-and-stick diagram because it uses rods and spheres to show the structure of the material. This example shows water, H_2O. There are two hydrogen atoms and one oxygen atom.

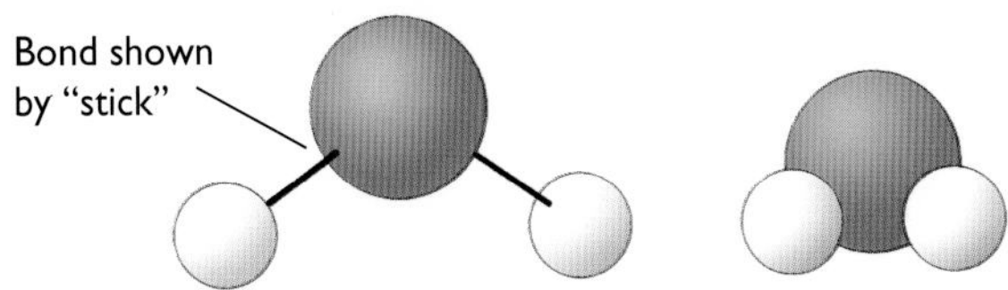

Colours too
The colours of each of the particles help differentiate the elements involved. The diagram can then be matched to the written and symbolic equation given with the diagram. In the case above, oxygen is red and hydrogen is grey.

Glossary of technical terms

absorb: to soak up a substance. Compare to adsorb.

acetone: a petroleum-based solvent.

acid: compounds containing hydrogen which can attack and dissolve many substances. Acids are described as weak or strong, dilute or concentrated, mineral or organic.

acidity: a general term for the strength of an acid in a solution.

acid rain: rain that is contaminated by acid gases such as sulphur dioxide and nitrogen oxides released by pollution.

adsorb/adsorption: to "collect" gas molecules or other particles on to the *surface* of a substance. They are not chemically combined and can be removed. (The process is called "adsorption".) Compare to absorb.

alchemy: the traditional "art" of working with chemicals that prevailed through the Middle Ages. One of the main challenges of alchemy was to make gold from lead. Alchemy faded away as scientific chemistry was developed in the 17th century.

alkali: a base in solution.

alkaline: the opposite of acidic. Alkalis are bases that dissolve, and alkaline materials are called basic materials. Solutions of alkalis have a pH greater than 7.0 because they contain relatively few hydrogen ions.

alloy: a mixture of a metal and various other elements.

alpha particle: a stable combination of two protons and two neutrons, which is ejected from the nucleus of a radioactive atom as it decays. An alpha particle is also the nucleus of the atom of helium. If it captures two electrons it can become a neutral helium atom.

amalgam: a liquid alloy of mercury with another metal.

amino acid: amino acids are organic compounds that are the building blocks for the proteins in the body.

amorphous: a solid in which the atoms are not arranged regularly (i.e. "glassy"). Compare with crystalline.

amphoteric: a metal that will react with both acids and alkalis.

anhydrous: a substance from which water has been removed by heating. Many hydrated salts are crystalline. When they are heated and the water is driven off, the material changes to an anhydrous powder.

anion: a negatively charged atom or group of atoms.

anode: the negative terminal of a battery or the positive electrode of an electrolysis cell.

anodising: a process that uses the effect of electrolysis to make a surface corrosion-resistant.

antacid: a common name for any compound that reacts with stomach acid to neutralise it.

antioxidant: a substance that prevents oxidation of some other substance.

aqueous: a solid dissolved in water. Usually used as "aqueous solution".

atom: the smallest particle of an element.

atomic number: the number of electrons or the number of protons in an atom.

atomised: broken up into a very fine mist. The term is used in connection with sprays and engine fuel systems.

aurora: the "northern lights" and "southern lights" that show as coloured bands of light in the night sky at high latitudes. They are associated with the way cosmic rays interact with oxygen and nitrogen in the air.

basalt: an igneous rock with a low proportion of silica (usually below 55%). It has microscopically small crystals.

base: a compound that may be soapy to the touch and that can react with an acid in water to form a salt and water.

battery: a series of electrochemical cells.

bauxite: an ore of aluminium, of which about half is aluminium oxide.

becquerel: a unit of radiation equal to one nuclear disintegration per second.

beta particle: a form of radiation in which electrons are emitted from an atom as the nucleus breaks down.

bleach: a substance that removes stains from materials either by oxidising or reducing the staining compound.

boiling point: the temperature at which a liquid boils, changing from a liquid to a gas.

bond: chemical bonding is either a transfer or sharing of electrons by two or more atoms. There are a number of types of chemical bond, some very strong (such as covalent bonds), others weak (such as hydrogen bonds). Chemical bonds form because the linked molecule is more stable than the unlinked atoms from which it formed. For example, the hydrogen molecule (H_2) is more stable than single atoms of hydrogen, which is why hydrogen gas is always found as molecules of two hydrogen atoms.

brass: a metal alloy principally of copper and zinc.

brazing: a form of soldering, in which brass is used as the joining metal.

brine: a solution of salt (sodium chloride) in water.

bronze: an alloy principally of copper and tin.

buffer: a chemistry term meaning a mixture of substances in solution that resists a change in the acidity or alkalinity of the solution.

capillary action: the tendency of a liquid to be sucked into small spaces, such as between objects and through narrow-pore tubes. The force to do this comes from surface tension.

catalyst: a substance that speeds up a chemical reaction but itself remains unaltered at the end of the reaction.

cathode: the positive terminal of a battery or the negative electrode of an electrolysis cell.

cathodic protection: the technique of making the object that is to be protected from corrosion into the cathode of a cell. For example, a material, such as steel, is protected by coupling it with a more reactive metal, such as magnesium. Steel forms the cathode and magnesium the anode. Zinc protects steel in the same way.

cation: a positively charged atom or group of atoms.

caustic: a substance that can cause burns if it touches the skin.

cell: a vessel containing two electrodes and an electrolyte that can act as an electrical conductor.

ceramic: a material based on clay minerals, which has been heated so that it has chemically hardened.

chalk: a pure form of calcium carbonate made of the crushed bodies of microscopic sea creatures, such as plankton and algae.

change of state: a change between one of the three states of matter, solid, liquid and gas.

chlorination: adding chlorine to a substance.

cladding: a surface sheet of material designed to protect other materials from corrosion.

clay: a microscopically small plate-like mineral that makes up the bulk of many soils. It has a sticky feel when wet.

combustion: the special case of oxidisation of a substance where a considerable amount of heat and usually light are given out. Combustion is often referred to as "burning".

compound: a chemical consisting of two or more elements chemically bonded together. Calcium atoms can combine with carbon atoms and oxygen atoms to make calcium carbonate, a compound of all three atoms.

condensation nuclei: microscopic particles of dust, salt and other materials suspended in the air, which attract water molecules.

conduction: (i) the exchange of heat (heat conduction) by contact with another object or (ii) allowing the flow of electrons (electrical conduction).

convection: the exchange of heat energy with the surroundings produced by the flow of a fluid due to being heated or cooled.

corrosion: the *slow* decay of a substance resulting from contact with gases and liquids in the environment. The term is often applied to metals. Rust is the corrosion of iron.

corrosive: a substance, either an acid or an alkali, that *rapidly* attacks a wide range of other substances.

cosmic rays: particles that fly through space and bombard all atoms on the Earth's surface. When they interact with the atmosphere they produce showers of secondary particles.

covalent bond: the most common form of strong chemical bonding, which occurs when two atoms *share* electrons.

cracking: breaking down complex molecules into simpler components. It is a term particularly used in oil refining.

crude oil: a chemical mixture of petroleum liquids. Crude oil forms the raw material for an oil refinery.

crystal: a substance that has grown freely so that it can develop external faces. Compare with crystalline, where the atoms are not free to form individual crystals and amorphous where the atoms are arranged irregularly.

crystalline: the organisation of atoms into a rigid "honeycomb-like" pattern without distinct crystal faces.

crystal systems: seven patterns or systems into which all of the world's crystals can be grouped. They are: cubic, hexagonal, rhombohedral, tetragonal, orthorhombic, monoclinic and triclinic.

cubic crystal system: groupings of crystals that look like cubes.

curie: a unit of radiation. The amount of radiation emitted by 1 g of radium each second. (The curie is equal to 37 billion becquerels.)

current: an electric current is produced by a flow of electrons through a conducting solid or ions through a conducting liquid.

decay (radioactive decay): the way that a radioactive element changes into another element decause of loss of mass through radiation. For example uranium decays (changes) to lead.

decompose: to break down a substance (for example by heat or with the aid of a catalyst) into simpler components. In such a chemical reaction only one substance is involved.

dehydration: the removal of water from a substance by heating it, placing it in a dry atmosphere, or through the action of a drying agent.

density: the mass per unit volume (e.g. g/cc).

desertification: a process whereby a soil is allowed to become degraded to a state in which crops can no longer grow, i.e. desert-like. Chemical desertification is usually the result of contamination with halides because of poor irrigation practices.

detergent: a petroleum-based chemical that removes dirt.

diaphragm: a semipermeable membrane – a kind of ultra-fine mesh filter – that will allow only small ions to pass through. It is used in the electrolysis of brine.

diffusion: the slow mixing of one substance with another until the two substances are evenly mixed.

digestive tract: the system of the body that forms the pathway for food and its waste products. It begins at the mouth and includes the stomach and the intestines.

dilute acid: an acid whose concentration has been reduced by a large proportion of water.

diode: a semiconducting device that allows an electric current to flow in only one direction.

disinfectant: a chemical that kills bacteria and other microorganisms.

dissociate: to break apart. In the case of acids it means to break up forming hydrogen ions. This is an example of ionisation. Strong acids dissociate completely. Weak acids are not completely ionised and a solution of a weak acid has a relatively low concentration of hydrogen ions.

dissolve: to break down a substance in a solution without a resultant reaction.

distillation: the process of separating mixtures by condensing the vapours through cooling.

doping: adding metal atoms to a region of silicon to make it semiconducting.

dye: a coloured substance that will stick to another substance, so that both appear coloured.

electrode: a conductor that forms one terminal of a cell.

electrolysis: an electrical–chemical process that uses an electric current to cause the break up of a compound and the movement of metal ions in a solution. The process happens in many natural situations (as for example in rusting) and is also commonly used in industry for purifying (refining) metals or for plating metal objects with a fine, even metal coating.

electrolyte: a solution that conducts electricity.

electron: a tiny, negatively charged particle that is part of an atom. The flow of electrons through a solid material such as a wire produces an electric current.

electroplating: depositing a thin layer of a metal onto the surface of another substance using electrolysis.

element: a substance that cannot be decomposed into simpler substances by chemical means

emulsion: tiny droplets of one substance dispersed in another. A common oil in water emulsion is milk. The tiny droplets in an emulsion tend to come together, so another stabilising substance is often needed to wrap the particles of grease and oil in a stable coat. Soaps and detergents are such agents. Photographic film is an example of a solid emulsion.

endothermic reaction: a reaction that takes heat from the surroundings. The reaction of carbon monoxide with a metal oxide is an example.

enzyme: organic catalysts in the form of proteins in the body that speed up chemical reactions. Every living cell contains hundreds of enzymes, which ensure that the processes of life continue. Should enzymes be made inoperative, such as through mercury poisoning, then death follows.

ester: organic compounds, formed by the reaction of an alcohol with an acid, which often have a fruity taste.

evaporation: the change of state of a liquid to a gas. Evaporation happens below the boiling point and is used as a method of separating out the materials in a solution.

exothermic reaction: a reaction that gives heat to the surroundings. Many oxidation reactions, for example, give out heat.

explosive: a substance which, when a shock is applied to it, decomposes very rapidly, releasing a very large amount of heat and creating a large volume of gases as a shock wave.

extrusion: forming a shape by pushing it through a die. For example, toothpaste is extruded through the cap (die) of the toothpaste tube.

fallout: radioactive particles that reach the ground from radioactive materials in the atmosphere.

fat: semi-solid energy-rich compounds derived from plants or animals and which are made of carbon, hydrogen and oxygen. Scientists call these esters.

feldspar: a mineral consisting of sheets of aluminium silicate. This is the mineral from which the clay in soils is made.

fertile: able to provide the nutrients needed for unrestricted plant growth.

filtration: the separation of a liquid from a solid using a membrane with small holes.

fission: the breakdown of the structure of an atom, popularly called "splitting the atom" because the atom is split into approximately two other nuclei. This is different from, for example, the small change that happens when radioactivity is emitted.

fixation of nitrogen: the processes that natural organisms, such as bacteria, use to turn the nitrogen of the air into ammonium compounds.

fixing: making solid and liquid nitrogen-containing compounds from nitrogen gas. The compounds that are formed can be used as fertilisers.

fluid: able to flow; either a liquid or a gas.

fluorescent: a substance that gives out visible light when struck by invisible waves such as ultraviolet rays.

flux: a material used to make it easier for a liquid to flow. A flux dissolves metal oxides and so prevents a metal from oxidising while being heated.

foam: a substance that is sufficiently gelatinous to be able to contain bubbles of gas. The gas bulks up the substance, making it behave as though it were semi-rigid.

fossil fuels: hydrocarbon compounds that have been formed from buried plant and animal remains. High pressures and temperatures lasting over millions of years are required. The fossil fuels are coal, oil and natural gas.

fraction: a group of similar components of a mixture. In the petroleum industry the light fractions of crude oil are those with the smallest molecules, while the medium and heavy fractions have larger molecules.

free radical: a very reactive atom or group with a "spare" electron.

freezing point: the temperature at which a substance changes from a liquid to a solid. It is the same temperature as the melting point.

fuel: a concentrated form of chemical energy. The main sources of fuels (called fossil fuels because they were formed by geological processes) are coal, crude oil and natural gas. Products include methane, propane and gasoline. The fuel for stars and space vehicles is hydrogen.

fuel rods: rods of uranium or other radioactive material used as a fuel in nuclear power stations.

fuming: an unstable liquid that gives off a gas. Very concentrated acid solutions are often fuming solutions.

fungicide: any chemical that is designed to kill fungi and control the spread of fungal spores.

fusion: combining atoms to form a heavier atom.

galvanising: applying a thin zinc coating to protect another metal.

gamma rays: waves of radiation produced as the nucleus of a radioactive element rearranges itself into a tighter cluster of protons and neutrons. Gamma rays carry enough energy to damage living cells.

gangue: the unwanted material in an ore.

gas: a form of matter in which the molecules form no definite shape and are free to move about to fill any vessel they are put in.

gelatinous: a term meaning made with water. Because a gelatinous precipitate is mostly water, it is of a similar density to water and will float or lie suspended in the liquid.

gelling agent: a semi-solid jelly-like substance.

gemstone: a wide range of minerals valued by people, both as crystals (such as emerald) and as decorative stones (such as agate). There is no single chemical formula for a gemstone.

glass: a transparent silicate without any crystal growth. It has a glassy lustre and breaks with a curved fracture. Note that some minerals have all these features and are therefore natural glasses. Household glass is a synthetic silicate.

glucose: the most common of the natural sugars. It occurs as the polymer known as cellulose, the fibre in plants. Starch is also a form of glucose. The breakdown of glucose provides the energy that animals need for life.

granite: an igneous rock with a high proportion of silica (usually over 65%). It has well-developed large crystals. The largest pink, grey or white crystals are feldspar.

Greenhouse Effect: an increase of the global air temperature as a result of heat released from burning fossil fuels being absorbed by carbon dioxide in the atmosphere.

gypsum: the name for calcium sulphate. It is commonly found as Plaster of Paris and wallboards.

half-life: the time it takes for the radiation coming from a sample of a radioactive element to decrease by half.

halide: a salt of one of the halogens (fluorine, chlorine, bromine and iodine).

halite: the mineral made of sodium chloride.

halogen: one of a group of elements including chlorine, bromine, iodine and fluorine.

heat-producing: see exothermic reaction.

high explosive: a form of explosive that will only work when it receives a shock from another explosive. High explosives are much more powerful than ordinary explosives. Gunpowder is not a high explosive.

hydrate: a solid compound in crystalline form that contains molecular water. Hydrates commonly form when a solution of a soluble salt is evaporated. The water that forms part of a hydrate crystal is known as the "water of crystallization". It can usually be removed by heating, leaving an anhydrous salt.

hydration: the absorption of water by a substance. Hydrated materials are not "wet" but remain firm, apparently dry, solids. In some cases, hydration makes the substance change colour, in many other cases there is no colour change, simply a change in volume.

hydrocarbon: a compound in which only hydrogen and carbon atoms are present. Most fuels are hydrocarbons, as is the simple plastic polyethene (known as polythene).

hydrogen bond: a type of attractive force that holds one molecule to another. It is one of the weaker forms of intermolecular attractive force.

hydrothermal: a process in which hot water is involved. It is usually used in the context of rock formation because hot water and other fluids sent outwards from liquid magmas are important carriers of metals and the minerals that form gemstones.

igneous rock: a rock that has solidified from molten rock, either volcanic lava on the Earth's surface or magma deep underground. In either case the rock develops a network of interlocking crystals.

incendiary: a substance designed to cause burning.

indicator: a substance or mixture of substances that change colour with acidity or alkalinity.

inert: nonreactive.

infra-red radiation: a form of light radiation where the wavelength of the waves is slightly longer than visible light. Most heat radiation is in the infra-red band.

insoluble: a substance that will not dissolve.

ion: an atom, or group of atoms, that has gained or lost one or more electrons and so developed an electrical charge. Ions behave differently from electrically neutral atoms and molecules. They can move in an electric field, and they can also bind strongly to solvent molecules such as water. Positively charged ions are called cations; negatively charged ions are called anions. Ions carry electrical current through solutions.

ionic bond: the form of bonding that occurs between two ions when the ions have opposite charges. Sodium cations bond with chloride anions to form common salt (NaCl) when a salty solution is evaporated. Ionic bonds are strong bonds except in the presence of a solvent.

ionise: to break up neutral molecules into oppositely charged ions or to convert atoms into ions by the loss of electrons.

ionisation: a process that creates ions.

irrigation: the application of water to fields to help plants grow during times when natural rainfall is sparse.

isotope: atoms that have the same number of protons in their nucleus, but which have different masses; for example, carbon-12 and carbon-14.

latent heat: the amount of heat that is absorbed or released during the process of changing state between gas, liquid or solid. For example, heat is absorbed when a substance melts and it is released again when the substance solidifies.

latex: (the Latin word for "liquid") a suspension of small polymer particles in water. The rubber that flows from a rubber tree is a natural latex. Some synthetic polymers are made as latexes, allowing polymerisation to take place in water.

lava: the material that flows from a volcano.

limestone: a form of calcium carbonate rock that is often formed of lime mud. Most limestones are light grey and have abundant fossils.

liquid: a form of matter that has a fixed volume but no fixed shape.

lode: a deposit in which a number of veins of a metal found close together.

lustre: the shininess of a substance.

magma: the molten rock that forms a balloon-shaped chamber in the rock below a volcano. It is fed by rock moving upwards from below the crust.

marble: a form of limestone that has been "baked" while deep inside mountains. This has caused the limestone to melt and reform into small interlocking crystals, making marble harder than limestone.

mass: the amount of matter in an object. In everyday use, the word weight is often used to mean mass.

melting point: the temperature at which a substance changes state from a solid to a liquid. It is the same as freezing point.

membrane: a thin flexible sheet. A semipermeable membrane has microscopic holes of a size that will selectively allow some ions and molecules to pass through but hold others back. It thus acts as a kind of sieve.

meniscus: the curved surface of a liquid that forms when it rises in a small bore, or capillary tube. The meniscus is convex (bulges upwards) for mercury and is concave (sags downwards) for water.

metal: a substance with a lustre, the ability to conduct heat and electricity and which is not brittle.

metallic bonding: a kind of bonding in which atoms reside in a "sea" of mobile electrons. This type of bonding allows metals to be good conductors and means that they are not brittle

metamorphic rock: formed either from igneous or sedimentary rocks, by heat and or pressure. Metamorphic rocks form deep inside mountains during periods of mountain building. They result from the remelting of rocks during which process crystals are able to grow. Metamorphic rocks often show signs of banding and partial melting.

micronutrient: an element that the body requires in small amounts. Another term is trace element.

mineral: a solid substance made of just one element or chemical compound. Calcite is a mineral because it consists only of calcium carbonate, halite is a mineral because it contains only sodium chloride, quartz is a mineral because it consists of only silicon dioxide.

mineral acid: an acid that does not contain carbon and that attacks minerals. Hydrochloric, sulphuric and nitric acids are the main mineral acids.

mineral-laden: a solution close to saturation.

mixture: a material that can be separated out into two or more substances using physical means.

molecule: a group of two or more atoms held together by chemical bonds.

monoclinic system: a grouping of crystals that look like double-ended chisel blades.

monomer: a building block of a larger chain molecule ("mono" means one, "mer" means part).

mordant: any chemical that allows dyes to stick to other substances.

native metal: a pure form of a metal, not combined as a compound. Native metal is more common in poorly reactive elements than in those that are very reactive.

neutralisation: the reaction of acids and bases to produce a salt and water. The reaction causes hydrogen from the acid and hydroxide from the base to be changed to water. For example, hydrochloric acid reacts with sodium hydroxide to form common salt and water. The term is more generally used for any reaction where the pH changes towards 7.0, which is the pH of a neutral solution.

neutron: a particle inside the nucleus of an atom that is neutral and has no charge.

noncombustible: a substance that will not burn.

noble metal: silver, gold, platinum, and mercury. These are the least reactive metals.

nuclear energy: the heat energy produced as part of the changes that take place in the core, or nucleus, of an element's atoms.

nuclear reactions: reactions that occur in the core, or nucleus of an atom.

nutrients: soluble ions that are essential to life.

octane: one of the substances contained in fuel.

ore: a rock containing enough of a useful substance to make mining it worthwhile.

organic acid: an acid containing carbon and hydrogen.

organic substance: a substance that contains carbon.

osmosis: a process where molecules of a liquid solvent move through a membrane (filter) from a region of low concentration to a region of high concentration of solute.

oxidation: a reaction in which the oxidising agent removes electrons. (Note that oxidising agents do not have to contain oxygen.)

oxide: a compound that includes oxygen and one other element.

oxidise: the process of gaining oxygen. This can be part of a controlled chemical reaction, or it can be the result of exposing a substance to the air, where oxidation (a form of corrosion) will occur slowly, perhaps over months or years.

oxidising agent: a substance that removes electrons from another substance (and therefore is itself reduced).

ozone: a form of oxygen whose molecules contain three atoms of oxygen. Ozone is regarded as a beneficial gas when high in the atmosphere because it blocks ultraviolet rays. It is a harmful gas when breathed in, so low level ozone, which is produced as part of city smog, is regarded as a form of pollution. The ozone layer is the uppermost part of the stratosphere.

pan: the name given to a shallow pond of liquid. Pans are mainly used for separating solutions by evaporation.

patina: a surface coating that develops on metals and protects them from further corrosion.

percolate: to move slowly through the pores of a rock.

period: a row in the Periodic Table.

Periodic Table: a chart organising elements by atomic number and chemical properties into groups and periods.

pesticide: any chemical that is designed to control pests (unwanted organisms) that are harmful to plants or animals.

petroleum: a natural mixture of a range of gases, liquids and solids derived from the decomposed remains of plants and animals.

pH: a measure of the hydrogen ion concentration in a liquid. Neutral is pH 7.0; numbers greater than this are alkaline, smaller numbers are acidic.

phosphor: any material that glows when energized by ultraviolet or electron beams such as in fluorescent tubes and cathode ray tubes. Phosphors, such as phosphorus, emit light after the source of excitation is cut off. This is why they glow in the dark. By contrast, fluorescors, such as fluorite, emit light only while they are being excited by ultraviolet light or an electron beam.

photon: a parcel of light energy.

photosynthesis: the process by which plants use the energy of the Sun to make the compounds they need for life. In photosynthesis, six molecules of carbon dioxide from the air combine with six molecules of water, forming one molecule of glucose (sugar) and releasing six molecules of oxygen back into the atmosphere.

pigment: any solid material used to give a liquid a colour.

placer deposit: a kind of ore body made of a sediment that contains fragments of gold ore eroded from a mother lode and transported by rivers and/or ocean currents.

plastic (material): a carbon-based material consisting of long chains (polymers) of simple molecules. The word plastic is commonly restricted to synthetic polymers.

plastic (property): a material is plastic if it can be made to change shape easily. Plastic materials will remain in the new shape. (Compare with elastic, a property where a material goes back to its original shape.)

plating: adding a thin coat of one material to another to make it resistant to corrosion.

playa: a dried-up lake bed that is covered with salt deposits. From the Spanish word for beach.

poison gas: a form of gas that is used intentionally to produce widespread injury and death. (Many gases are poisonous, which is why many chemical reactions are performed in laboratory fume chambers, but they are a byproduct of a reaction and not intended to cause harm.)

polymer: a compound that is made of long chains by combining molecules (called monomers) as repeating units. ("Poly" means many, "mer" means part).

polymerisation: a chemical reaction in which large numbers of similar molecules arrange themselves into large molecules, usually long chains. This process usually happens when there is a suitable catalyst present. For example, ethene reacts to form polythene in the presence of certain catalysts.

porous: a material containing many small holes or cracks. Quite often the pores are connected, and liquids, such as water or oil, can move through them.

precious metal: silver, gold, platinum, iridium, and palladium. Each is prized for its rarity. This category is the equivalent of precious stones, or gemstones, for minerals.

precipitate: tiny solid particles formed as a result of a chemical reaction between two liquids or gases.

preservative: a substance that prevents the natural organic decay processes from occurring. Many substances can be used safely for this purpose, including sulphites and nitrogen gas.

product: a substance produced by a chemical reaction.

protein: molecules that help to build tissue and bone and therefore make new body cells. Proteins contain amino acids.

proton: a positively charged particle in the nucleus of an atom that balances out the charge of the surrounding electrons

pyrite: "mineral of fire". This name comes from the fact that pyrite (iron sulphide) will give off sparks if struck with a stone.

pyrometallurgy: refining a metal from its ore using heat. A blast furnace or smelter is the main equipment used.

radiation: the exchange of energy with the surroundings through the transmission of waves or particles of energy. Radiation is a form of energy transfer that can happen through space; no intervening medium is required (as would be the case for conduction and convection).

radioactive: a material that emits radiation or particles from the nucleus of its atoms.

radioactive decay: a change in a radioactive element due to loss of mass through radiation. For example uranium decays (changes) to lead.

radioisotope: a shortened version of the phrase radioactive isotope.

radiotracer: a radioactive isotope that is added to a stable, nonradioactive material in order to trace how it moves and its concentration.

reaction: the recombination of two substances using parts of each substance to produce new substances.

reactivity: the tendency of a substance to react with other substances. The term is most widely used in comparing the reactivity of metals. Metals are arranged in a reactivity series.

reagent: a starting material for a reaction.

recycling: the reuse of a material to save the time and energy required to extract new material from the Earth and to conserve non-renewable resources.

redox reaction: a reaction that involves reduction and oxidation.

reducing agent: a substance that gives electrons to another substance. Carbon monoxide is a reducing agent when passed over copper oxide, turning it to copper and producing carbon dioxide gas. Similarly, iron oxide is reduced to iron in a blast furnace. Sulphur dioxide is a reducing agent, used for bleaching bread.

reduction: the removal of oxygen from a substance. See also: oxidation.

refining: separating a mixture into the simpler substances of which it is made. In the case of a rock, it means the extraction of the metal that is mixed up in the rock. In the case of oil it means separating out the fractions of which it is made.

refractive index: the property of a transparent material that controls the angle at which total internal reflection will occur. The greater the refractive index, the more reflective the material will be.

resin: natural or synthetic polymers that can be moulded into solid objects or spun into thread.

rust: the corrosion of iron and steel.

saline: a solution in which most of the dissolved matter is sodium chloride (common salt).

salinisation: the concentration of salts, especially sodium chloride, in the upper layers of a soil due to poor methods of irrigation.

salts: compounds, often involving a metal, that are the reaction products of acids and bases. (Note "salt" is also the common word for sodium chloride, common salt or table salt.)

saponification: the term for a reaction between a fat and a base that produces a soap.

saturated: a state where a liquid can hold no more of a substance. If any more of the substance is added, it will not dissolve.

saturated solution: a solution that holds the maximum possible amount of dissolved material. The amount of material in solution varies with the temperature; cold solutions can hold less dissolved solid material than hot solutions. Gases are more soluble in cold liquids than hot liquids.

sediment: material that settles out at the bottom of a liquid when it is still.

semiconductor: a material of intermediate conductivity. Semiconductor devices often use silicon when they are made as part of diodes, transistors or integrated circuits.

semipermeable membrane: a thin (membrane) of material that acts as a fine sieve, allowing small molecules to pass, but holding large molecules back.

silicate: a compound containing silicon and oxygen (known as silica).

sintering: a process that happens at moderately high temperatures in some compounds. Grains begin to fuse together even through they do not melt. The most widespread example of sintering happens during the firing of clays to make ceramics.

slag: a mixture of substances that are waste products of a furnace. Most slags are composed mainly of silicates.

smelting: roasting a substance in order to extract the metal contained in it.

smog: a mixture of smoke and fog. The term is used to describe city fogs in which there is a large proportion of particulate matter (tiny pieces of carbon from exhausts) and also a high concentration of sulphur and nitrogen gases and probably ozone.

soldering: joining together two pieces of metal using solder, an alloy with a low melting point.

solid: a form of matter where a substance has a definite shape.

soluble: a substance that will readily dissolve in a solvent.

solute: the substance that dissolves in a solution (e.g. sodium chloride in salt water).

solution: a mixture of a liquid and at least one other substance (e.g. salt water). Mixtures can be separated out by physical means, for example by evaporation and cooling.

solvent: the main substance in a solution (e.g. water in salt water).

spontaneous combustion: the effect of a very reactive material beginning to oxidise very quickly and bursting into flame.

stable: able to exist without changing into another substance.

stratosphere: the part of the Earth's atmosphere that lies immediately above the region in which clouds form. It occurs between 12 and 50 km above the Earth's surface.

strong acid: an acid that has completely dissociated (ionised) in water. Mineral acids are strong acids.

sublimation: the change of a substance from solid to gas, or vica versa, without going through a liquid phase.

substance: a type of material, including mixtures.

sulphate: a compound that includes sulphur and oxygen, for example, calcium sulphate or gypsum.

sulphide: a sulphur compound that contains no oxygen.

sulphite: a sulphur compound that contains less oxygen than a sulphate.

surface tension: the force that operates on the surface of a liquid, which makes it act as though it were covered with an invisible elastic film.

suspension: tiny particles suspended in a liquid.

synthetic: does not occur naturally, but has to be manufactured.

tarnish: a coating that develops as a result of the reaction between a metal and substances in the air. The most common form of tarnishing is a very thin transparent oxide coating.

thermonuclear reactions: reactions that occur within atoms due to fusion, releasing an immensely concentrated amount of energy.

thermoplastic: a plastic that will soften, can repeatedly be moulded it into shape on heating and will set into the moulded shape as it cools.

thermoset: a plastic that will set into a moulded shape as it cools, but which cannot be made soft by reheating.

titration: a process of dripping one liquid into another in order to find out the amount needed to cause a neutral solution. An indicator is used to signal change.

toxic: poisonous enough to cause death.

translucent: almost transparent.

transmutation: the change of one element into another.

vapour: the gaseous form of a substance that is normally a liquid. For example, water vapour is the gaseous form of liquid water.

vein: a mineral deposit different from, and usually cutting across, the surrounding rocks. Most mineral and metal-bearing veins are deposits filling fractures. The veins were filled by hot, mineral-rich waters rising upwards from liquid volcanic magma. They are important sources of many metals, such as silver and gold, and also minerals such as gemstones. Veins are usually narrow, and were best suited to hand-mining. They are less exploited in the modern machine age.

viscous: slow moving, syrupy. A liquid that has a low viscosity is said to be mobile.

vitreous: glass-like.

volatile: readily forms a gas.

vulcanisation: forming cross-links between polymer chains to increase the strength of the whole polymer. Rubbers are vulcanised using sulphur when making tyres and other strong materials.

weak acid: an acid that has only partly dissociated (ionised) in water. Most organic acids are weak acids.

weather: a term used by Earth scientists and derived from "weathering", meaning to react with water and gases of the environment.

weathering: the slow natural processes that break down rocks and reduce them to small fragments either by mechanical or chemical means.

welding: fusing two pieces of metal together using heat.

X-rays: a form of very short wave radiation.

Index